David Jones

Diseases of the bladder and prostate, and obscure affections of the urinary organs

with diagrams illustrating the author's treatment of stone, without cutting, and

numerous successfully treated cases with the spray treatment hitherto

David Jones

Diseases of the bladder and prostate, and obscure affections of the urinary organs
with diagrams illustrating the author's treatment of stone, without cutting, and numerous
successfully treated cases with the spray treatment hitherto

ISBN/EAN: 9783744736268

Printed in Europe, USA, Canada, Australia, Japan

Cover: Foto ©berggeist007 / pixelio.de

More available books at **www.hansebooks.com**

DISEASES

OF THE

Bladder and Prostate,

AND OBSCURE AFFECTIONS OF THE

URINARY ORGANS

With Diagrams

ILLUSTRATING THE AUTHOR'S TREATMENT OF

STONE,

WITHOUT CUTTING,

AND NUMEROUS SUCCESSFULLY TREATED

CASES

WITH THE SPRAY TREATMENT

HITHERTO REGARDED AS "INCURABLE."

BY

DAVID JONES, M.D.

SIXTH EDITION.

LONDON:
SIMPKIN, MARSHALL & CO., STATIONERS HALL COURT.
C. MITCHELL & CO., RED LION COURT, FLEET ST.
1890.

NOTICE.

To facilitate the treatment, and to accelerate the cure, of cases such as are described in these pages, and of other cases of a serious character, Dr. DAVID JONES will be happy, if requested to do so, to forward addresses of suitable private apartments where patients may be comfortably accommodated whilst under treatment.

Full particulars may be obtained on application to Dr. JONES, 15, Welbeck Street, Cavendish Square, W., at which place he may be consulted daily from 11 till 1, *(Tuesdays and Fridays by appointment only)*.

To prevent disappointment and inconvenience, patients residing at a distance, and wishing to consult DR. JONES, had better make an appointment before coming.

PREFACE.

TO

SIXTH EDITION.

————

SIX years and more have elapsed since the issue of the fifth edition of this work, and in the interval nearly six thousand copies have been sold. The author has reason to be grateful for the public appreciation he has received, and the more that the demand continues. There are sufficient reasons why that edition should be supplanted by another. The particular department of surgical science under consideration is subject to continual development; and readers are entitled at reasonable periods to be posted in newer theories and more recent discoveries. Much that is of interest has happened. Electrolysis, for example, long regarded with distrust and repressed to the background, has begun to win upon confidence and made some strides to the front. Some reference to its position as a curative agent as applied to urethral stricture will be found in these pages. Time moreover has been afforded for a right appreciation of Professor BIGELOW's developments in the treatment of stone described in the fifth edition (pp. 40—48). The writer is enabled to give the results of his experience gained in hundreds of cases of the professor's method of removing stone in young children and is happy to say that it has proved highly successful. With regard to the professor's operation in substitution of lithotrity (as it used to be practised) no one can doubt of its triumph, nor could any justification be

*2

now pleaded for removing stone by a succession of operations and allowing the fragments to tear the urethra in their exit, when by a single operation, though somewhat more prolonged, the end can be attained by BIGELOW's tubes without injuring the urethra at all. The writer in the previous edition expressed a belief that save in some exceptional cases BIGELOW's method moreover would altogether supplant lithotomy (the cutting operation), and lithotrity according to CIVIALE's plan, and at the same time make the removal of stone the most successful operation in modern surgery. When surgeons in general become better diagnostitians there is little doubt that "litholapaxy" will supersede all other methods for the purpose.

Such are some notable instances of advance in the department of urinary surgery. But the record of blunders goes on steadily swelling. Among the more flagrant is the case of a lady (reported on page 202) who had suffered for thirty-three years from stone in the bladder, but though resorting to the most eminent authorities was treated for paralysis of the organ and where the writer's son (GORDON G. JONES) assisted by the writer himself, removed a calculus weighing over three ounces and measuring fifty millimetres by forty, or in English measure two inches and one-eighth by one inch and three-quarters.

Opportunity has not served the writer's intention in publishing this sixth edition to make the work more comprehensive, with some notice of *medical* diseases of the genito-urinary system, including Bright's disease of the kidneys, diabetes and various nervous affections (so called) so successfully treated by LALLEMANDE (in France), and the late Dr. DAWSON (in London). He will remark however in passing, since the question has been raised, that for what is sometimes called "hypogastria in the male," but Sir JAMES PAGET (in his excellent lecture on the subject) terms "sexual hypochondriasis" and is vernacularly known as "spermatorrhœa," while he is led to protest most emphatically against cauterisation, he at the same time fully concurs in the application of local means which is in accordance with his own general views.

In order to make the book more useful in point of reference the writer has added, at the suggestion of some of his patients, an epitomised index of the published cases, that the reader may discern at a glance the nature of what is recorded. Further, for purposes of verification, names and addresses are added. The writer, indeed, is unacquainted with any medical work affording so much facility for direct information or so much assurance of good faith.

Special attention may be called to the class of cases—where there is no stone,—which are reported obscure and incurable. Among these is disease of the prostate, declared by high authorities* to be " incurable by any known means." The writer begs to refer in contradiction to this statement, to the numerous cases of cure recorded in this book. It is true, as alleged, that some of the patients have eventually died. The allegation has been dealt with before (in the second edition of " Urinary Diseases " and elsewhere) ; but generally the end has come from some other disease or else at a ripe old age. The Rev. CANON COCKIN, for instance, who about a year ago succumbed to paralysis, aged 76, but who nevertheless enjoyed nine years of freedom from the affliction of which he was cured.

With respect to the special means of cure which the writer possesses in these and some other classes of cases, the profession asks, as has been remarked upon in former issues, why not, in the interests of the afflicted, make them known. The answer is that, as in other departments of industry, a discoverer is entitled to the fruits of his industry and research. Will the profession, which has the good of mankind so much at heart, co-operate to buy the discovery ? Certainly they will not. They will only make reproach that it is not surrendered for nothing. The writer is

---

* See article on " diseased prostate " (page 56).

willing to yield something. He offers his little hospital in Dean Street, with his services for life, to any one who will endow it. He is moreover authorized by his son, GORDON G. JONES, to say that he, in like manner heartily co-operates in the much needed object. Of his service to the helpless from time to time, he may not speak. Why must he, more than other men, be made up of sacrifice, and ignore the claims of his family? Success in professional life is apt to bring something besides honour. There are those to whom it is the most unpardonable of offences, the one thing which cannot be forgiven. It is not at all certain that on following the advice of the envious, the writer ought not to incur their derision. As the lesser of two evils, he prefers to accept their enmity, manifested sometimes in execrable sentiment and more execrable taste.

PREFACE

FIFTH EDITION.

———

THIS fifth edition, though a considerable enlargement of former issues, is not presented as, by any means, an exhaustive treatise upon the subjects to which it relates. The writer's chief object has been to present some typical cases illustrative of his method of treatment of stone, of diseases of the bladder generally, and of the prostate gland, and so to describe the symptoms of each case that sufferers, recognising therein symptoms similar to their own, may be sustained in their hope of cure, notwithstanding other and previously applied means or methods may have disappointed that hope—and the writer believes that as auxiliary to this, the information given in the first four Parts of this present edition will be found very useful to lay readers, and, possibly, to some members of the profession also. Indeed, the information embodied in those parts elucidates the general subject, and, to some extent, the writer's method of treatment likewise.

Experience, in the active pursuit of his profession, for nearly thirty years, has convinced the writer that popular ignorance is productive of much avoidable suffering, and that the public ought, therefore, to be furnished with information respecting the diseases in question—diseases so distressing and so common. Moreover, the public are the patients, and, consequently, the persons most concerned in the matter, and the press is the chief medium of communication with them. The "orthodox" view may be, that books on medical subjects are to be read only by medical men, or by those who are intending to become such—though everybody knows that it has now become a "rage" to acquire knowledge, and to read and study medical works as well as philosophical writings generally. At all events medical books when *published* are just as much books accessible to the public as this present little work can be, and the authors of them, even such eminent authors as SIR JAMES PAGET and SIR HENRY THOMPSON, would, probably, not be very greatly displeased if the public demand for their works were such as to create very frequent necessity for further editions—the public becoming purchasers of their books as well as patients for their services   In noticing the recent edition of a work by the last-named author, the "Lancet" (July 8th, 1882—p. 12) says :—"The author has made a new departure in the form in which this edition is published. Following a course which has been recently pursued in works of general literature he has issued it at less than a fourth of its former price." Doubtless the author's motive was a very commendable one, viz., to bring the edition in question within reach of even poor students ; still, as a matter of fact the work is, as a published book, thus brought within reach of the general public also—and who shall say that SIR ERASMUS WILSON's published works on Diseases of the Skin, including his "Popular" Treatises, have not brought both himself and his knowledge and skill, as a specialist in dermatology, under the notice of the public, as well as of the medical profession, and thus, indirectly at least, contributed to his professional success? Many other, and similar instances might be mentioned. "New Methods" and "New Discoveries" are, in fact, mentioned and advertised in medical and other journals every day. The writer has, however, special reasons for preparing and publishing this little book in its present form. One reason has been already alluded to, and has reference to patients themselves—and if thus communicating with the public needs justification, such justification is found in the fact (provable by unquestionable testimony) that, in numerous instances, the really incurable condition of some patients, eventually resorting to the writer, has been traceable to the delay caused by the persuasions brought to bear upon them not to consult him—a circumstance sufficient in itself to force any man, possessing any degree of manly courage, paternal affection, and humane feeling, to the resolve, not to allow any skill with which he may be gifted, to remain unused or crushed beneath the heel of a professional despotism. Other reasons concern both the writer's method of treatment, and the question of divulgence of that method. These will now be more particularly referred to.

The writer claims for his treatment of Stone, and other Diseases of the Bladder and associated organs, the application of a NEW Method, and even without the fate of an eminent professor confronting him, he would have avoided any premature disclosure of it, and the risk of its being experimented with by unpractised hands, and before even the writer himself shall have demonstrated its applicability to *all* the various forms and phases (so to speak) of the diseases in question. Professor CLAY, of Birmingham, a surgeon of acknowledged repute, some time ago published in the medical journals a Discovery for the Cure of Cancer by Chian Turpentine. Cases which he had cured were published[*] and they were both remarkable and interesting. They were undoubtedly genuine, and, in Professor CLAY'S hands, had been, as undoubtedly, successfully cured. The soothing, narcotic, and curative effects of the therapeutic agent discovered were shewn with such force as to attract the surprised attention of the whole

---

[*] See "Lancet," March 27th, 1880—p. 477, "On the treatment of cancer of the female generative organs by a new method."

profession. The history of the cases—the gradual return to health of the patients—the ocular demonstration of healthy granulations in place of malignant growths—all led to the revealed discovery being put to the test of a general application by general practitioners and others—the result being, however, most unfavorable—Professor CLAY himself (a man of undoubted skill in medicine and surgery, and of unimpeachable character), having to share in the obloquy heaped on his discovery, all which would, probably, have been avoided if the professor had kept secret both his discovery and his method of applying it for a little while longer, and until more experience, covering a more extended area of observation, in treatment of cases, had guided him in his conclusions. Such a course might have treasured up indisputable proof that Chian Turpentine *is* a therapeutic agent of great value, and have afforded such full and precise specification of the *modus operandi* to be employed, as could scarcely have been unsuccessful, even in comparatively unskilful hands, thus contributing to results equally honoring to the discoverer and beneficial to the public. Possibly, the professor might, meanwhile, have been severely censured for keeping his discovery secret, though intending to keep it so only for a time, and perhaps denounced as acting " unprofessionally." Events might, however, have justified his patient endurance, and ensured real success to all concerned.[*]

These remarks will not, of course be understood as implying, on the part of the writer, any opinion, either favorable or unfavorable, respecting Professor CLAY's Discovery, but only as indicating his belief that the unfavorable results which followed the divulgence of it may have been owing chiefly to the

---

[*] It may be added that, notwithstanding the unfavorable result above alluded to, Professor CLAY has himself publicly re-affirmed his own conviction of the therapeutic value of his discovery. He says:—" I have nothing to withdraw or to qualify as regards the statements I made (as the result of observation) as to the effects of Chian Turpentine in uterine cancer." See " Lancet," December 17th, 1881—p. 1033, " On the use of Chian Turpentine in cancer.

circumstance that it was prematurely placed within the reach of unpractised, unskilful, and, it may be, jealous hands.

The foregoing observations indicate, however, pretty clearly the reasons which influence the writer in his resolve to decline, for the present, fully to reveal his method of treatment of stone and other diseases of the bladder (male and female) and of the prostate gland, as respects both the medicaments used, and the surgical appliances adopted. Not that any present reticence on his part implies any lack of confidence in the soundness of the principles on which his treatment is based. That treatment has been remarkably successful now for many years, and, in numerous instances, where the cases coming under his observation had been pronounced "hopelessly incurable" by some of the most eminent specialists of the day. His success has, naturally enough, excited some attention, and a desire for full and precise information as to the *modus operandi* employed, and so long as such information is withheld, the writer will, no doubt, expose himself to the taunt— "You are practising a secret, and that is unprofessional,"—less unprofessional, however, and less censurable too, the writer ventures to think, than to *profess* to give publicity to a new discovery, and secretly to practise the essential principle of it notwithstanding, as some have done. Alluding to the writer's present reticence, an eminent surgeon, and late President of the Royal College of Surgeons, writing from the Continent, says :—"I have received your publication—you do not disclose your discovery, you do not tell us anything. The profession will not meet you unless you explain your mode of treatment."

Well, notwithstanding long and patient thought and attention to the study of the subject, notwithstanding also a very extensive and successful application in actual practise, of the results of that study, the writer feels that still further experience is necessary, or at least desirable, before pronouncing the practical appliances invented and used as absolutely perfect, or determining that the method of treatment employed is absolutely suitable to *every* class of disease of the bladder and prostate,—consequently, before the *modus operandi* shall be presented and explained to the general body of medical practitioners.

In further support of this cautious procedure the writer has, he thinks, only to remind the more skilful in the profession that all medical practitioners have not the "*tactus eruditus*" which some have, and that the dexterity of manipulation *acquired* by one, by long experience, and careful, skilful practice, cannot be *imparted* to others. The writer, moreover, firmly believes that very little, comparatively, is known by medical practitioners in general respecting diseases of the bladder and prostate; and that few, very few, of them, have made those diseases their special study. Specialists who have done so know full well that lamentable ignorance respecting those diseases prevails among such prac-titioners. They know, too, that one man handles a catheter or "cuts" for stone far better than another, and when the extreme sensitiveness of patients who suffer from bladder diseases is considered it will be obvious that the greatest possible delicacy is requisite in dealing with such cases, and that any clumsiness or roughness would be positively dangerous. Not long since a patient in the neighbourhood of the writer's residence, died from the effects of rupture of the urethra behind a stricture; another was ruined for life by a clumsy practitioner endeavouring to "*force*" a catheter into the bladder. Surgical literature furnishes abundant proof bearing on this point.* The truth is that physical force will not do in cases where the bladder has to be dealt with. To proceed with any reasonable hope of success, both the bladder and urethra must be treated as one would treat a troublesome lock—gentleness may succeed, force is sure to fail. In most cases *force* is rough, unscientific, uncertain, and dangerous. A young woman once

---

* Notwithstanding the opinions expressed in medical journals adverse to special hospitals, the writer feels confident that such institutions furnish the best means by which perfection may be attained in any one particular branch of medicine or surgery.

came under the writer's care who, treated by the physical force method, had been literally lacerated from the external meatus to the vesical sphincter, in other words, from the entrance of the urethra to the neck of the bladder. She had extensive cystitis (inflammation of the bladder) had quite lost the power of retaining her urine, and was, indeed, as pitiable an object as could well be conceived. The writer could mention, or refer to, several other cases of a similarly deplorable character. One other case only may, however be alluded to, and that is the case of the wife of a physician (holding an important position in connection with medical journalism) whose case was by him committed to the care of the writer, who from the first, pronounced it really incurable, and solely in consequence of the " forcible and rapid dilatation " previously applied in the treatment of the case—a conclusion in which the husband (the physician alluded to) fully concurred. The writer does not hesitate to say that this barbarous treatment— " forcible and rapid dilatation " of the female urethra—has rendered many cases incurable by any mode of treatment. Two cases, one from Leeds, another from Newbury, which came under his notice were rendered incurable and miserable for life by the above method of treatment. He remarked to both patients that he could cure disease, but not effects of bad surgery, and be it remembered the two cases alluded to were not lacerated as in the *former* cases, but forcibly and rapidly dilated. These considerations, with others which might be urged, shew the importance of committing the power to operate, in these cases, only to persons who have made such diseases their *special* study, and whose skill and experience qualify them for the work.

On these grounds, therefore, the writer declines for the present, fully to explain the *modus operandi* of his treatment of diseases of the bladder and prostate. Both his reasons, and his mode of giving effect to them, may, possibly, meet with disapproval at the hands of his professional brethren —yes, "professional brethren," for, whether acknowledged or repudiated, a professional relationship does, in fact, exist between him and the other members of the medical profession. He possesses the friendship of many, and has, he believes, not one *personal* enemy among them— nay, more, has their confidence. He has treated the wives and families of several medical men, and has successfully performed operations for stone and other diseases on medical men *themselves*, both allopathic and homœopathic. Any *animus*, if it exist, is, he is persuaded, purely professional, and attributable to that jealousy, which, unfortunately, obtains, not in the profession of medicine alone, but in other professions also, especially manifesting itself where one man's superior skill happens to be applied with signal success where another's skill of lower degree, has failed.* Still, to represent the writer as a "quack" and "a man who possesses no medical or surgical qualification whatever" is simply absurd, and *absolutely false*—although even then there are patients, in large numbers, who would willingly say of him :— "Whether he be a quack or no, we know not, one thing we know, and that is, that whereas we were once diseased, and by others pronounced 'incurable,' Dr. Jones has cured us." The medical profession, know, however, full well, that the writer is not an untaught, untrained, pretender to medical and surgical knowledge and skill, but that, on the contrary, he has had an education far more extensive, and complete, both at home and abroad, than is ordinarily required as qualifying for the practice of medicine and surgery, as is evidenced by his possession of the following diplomas, obtained after examinations searching and severe, viz. :—

1. Royal College of Surgeons† (London) ... 20th August, 1847.

*Not that the writer is narrowly opposed to the well intentioned exercise of any skill that may be possesed by others, or even to the well intentioned application of either allopathic or homœopathic views—as is evidenced by his willingness to allow the Home Hospital which he has founded in Dean Street, Soho, to be used, as it now is, by two practitioners entertaining those views.

†For this diploma Dr. JONES passed his examinaton in the name of David Griffiths Jones—the name Griffiths (a family name) having then been *adopted* (by usage from his childhood) and continued for some years in addition to his proper baptismal name David.

2. Royal College of Physicians (London) ... 17th July, 1865.

3. University of Heidelberg Degree of M.D. ⎫
   (Summa cum laude) ... ... ⎬ 25th March, 1865.

Dr. JONES also passed his examination for ⎫
   the Degree of M.D. in the University ⎬ in April, 1865.
   of St. Andrews ... ... ⎭

The public, too, know that a duly obtained diploma, whether registered or not, is, by the law, deemed a sufficient proof of the right and power of its possessor to practise medicine and surgery. An eminent judge, in a case which not long since came before him, stated that many such unregistered medical practitioners were among his own personal friends — and if, in the face of these facts, misrepresentation against the writer, shall be further indulged in, the public will know how to appreciate it, and to what to attribute it. The writer feels, moreover, that, under the circumstances, he is fully justified in applying any skill he may possess, in relation to the treatment of the diseases, the subject of this present work, as well for the benefit of himself and family, as for that of the public. The profession of medicine is, no doubt, a humane profession, and its members are deservedly credited with humanity, and benevolence too, in their pursuit of it, and of this credit the writer can honestly claim a share, for he has gratuitously treated and cured numberless cases coming under his care—indeed, two Free Beds have been appropriated by him, and often filled, in the "Home Hospital" already referred to.† Nor would he for one moment, wish that such philanthropy should be either disallowed or discouraged. But a theory may foster a monopoly, and in the face of many theoretic usages, inimical as well to the progress of the science of medicine itself as to the interests of individual members of the profession, every day practice declares, and very emphatically too,

*The first *summa cum laude* Degree (a Degree of the highest possible praise) ever conferred by that University upon a foreigner (an Englishman).

† For further particulars respecting the Home Hospital, see the announcement at the end of this book.

that medical men, like other men, do, in fact, labor to support themselves and those dependent upon their exertions. As in other professions, so in the medical, and chiefly through the working of monopoly-tending theories and usages, few only realize large, or even remunerative incomes. The majority, incessant in toil, and strangers to domestic comfort, can seldom retire from their work to spend, in quiet, any " remaining days " they may chance to have-— on the contrary, they more usually work like slaves, die like cab-horses, and leave their families dependent upon the kindness, it may be the charity, of others. This would not be, to any such extent as it is, if the principle that " the labourer is worthy of his hire " were allowed a plain, common-sense, truthful recognition, and a free, full application, in the medical profession, as in other spheres of human thought, skill, and industry, and if medical men were not, by falsely-grounded " professional" usages *supposed* self-sacrificingly to *give* for the benefit of "suffering humanity " the special services of their brain and of their hands. The writer cares not, however, to urge these latter considerations in vindication of his resolve not yet fully to publish and explain his method of treatment, as illustrated and justified by the cases recorded in this book. He is content to base that resolve on other grounds— especially on the conviction that still further experience is desirable, and which experience must come from or through those who, deriving useful information herefrom, may resort to him for treatment.

When this is ensured, and the writer has proven, to still further demonstration, the applicability of his method of treatment to *all* classes of disease of the bladder and prostate, and when professional *animus* shall have ceased to display itself, the writer will then be prepared fully to disclose his discovery, and his method of treatment of the diseases described in this present work—a method which has already been signally successful where other, and more ordinary, methods have as signally failed.

It only remains to add, with respect to the cases recorded in this present edition (specimens only of a very large number of similar cases), that no one can reasonably claim the right to question their genuineness, who shall not have previously sought verification of them. Such verification is readily accessible to all enquirers. In the Appendix the full names and addressess of many of the patients are (by permission) set forth, and in almost every other case the writer is at liberty to furnish the name and address, and he will willingly do so on application being made to him for the same.

The names of the medical men who had, in almost every instance, attended the cases, previously to their coming under the writer's hands, are very numerous, and if appended, would have shown, at a glance, that the ordinary methods of treatment had been applied by (amongst others) the most eminent specialists of the day. On the present occasion, however, such names are not appended— but they may, if desired, be ascertained, in any particular case or cases, by communicating with the patients themselves.

INDEX.

NOTICE.

The index is so arranged, that an outline is given in it
of every case reported; the reader can therefore see at a glance
if it be desirable to read the case in detail, as given the text.

INDEX.

PAGE.

*Urinary Diseases : analysis of 500 cases, &c., By DAVID JONES, M.D.
 Cases No. 9 & 10, edited by GORDON G. JONES, second edition, Simpkin,
 Marshall & Co., C. Mitchell & Co.

|See cases in writer's last book : Homœopathy, its truth, its law of cure,
 and its statistics, pages 86 to 91. [To be had from the Secretary, Home
 Hospital, 10, Dean Street, Soho.]

Thomas' Hospital, and St. Peter's Hospital for
stone—continued well for ten years and died of
bronchitis some time ago, 75 years old ... 91

Case No. 7.—E. B.

BLADDER DISEASE WITH STONE. The patient
came to the writer 16 years ago with very severe
symptoms. Owing to an extremely contracted
urethral opening, only the smallest possible
instruments could be used on which account he
could not be "sounded"—some time after he
was cured the patient called on the writer, he
and his assistant, with the aid of a small sound
detected stone, the patient however did not
believe it as he continued well; he subsequently
married and became father of three children; he
refuses even now to believe he has stone, but he
certainly has one 95

Case (unnumbered).—I. C. W. I.

BLADDER DISEASE complicated with Stone.
The case had been treated as for "chronic
prostatitis" only. The patient came to the writer
in a deplorable condition, having got no relief from
anything but morphia, which, he added, was fast
taking away his life.—Obtained relief from first
"spray" which liberated a stone, which was afterwards removed to the great comfort of the patient;
another large stone was too deeply impacted for
removal. He lived in comparative ease for a long
while, but died eventually of kidney complications
and senile decay 99

Case No. 8.—R. C., aged 29, single.

BLADDER DISEASE of a chronic and obscure nature
complicated with prostatic trouble which none of his

hospital and came to Dean Street—great difficulty
was experienced in his treatment and the usual
means doing no service another method was
adopted which proved efficacious—ultimately he
was operated upon and successfully treated by
gentle means, but the treatment occupied nearly
three months instead of the usual 14 to 21 days ...

Case No. 25.—S. H. T., aged 35, single.

BLADDER DISEASE accompanied with paralysis ;
unsuccessfully attended the Birmingham Free
Hospital, and from time to time was treated by
ten allopathic physicians and surgeons without
relief—the Birmingham Homœopathic Hospital
was now tried for two months with no good result
—galvanism and other means were employed
till one of the surgeons told him his case was
" hopeless " and he was ultimately discharged as
" incurable."—This case rapidly responded to the
" spray " treatment,—the patient has continued
well ever since, and has held the post of engineer
in the South Yorkshire Lunatic Asylum, Wadsley,
near Sheffield ... ...

Case No. 26.—S. M.

BLADDER DISEASE and ENLARGED PROSTATE.—
Catheter wholly discontinued after being continuously used for upwards of two years—the
patient is now 80 years' old and enjoys fair health
for his time of life ... ...

Case No. 27.—H. D.

BLADDER PARALYSIS.—This was a very advanced
case attended with constant dribbling of urine
night and day,—the patient had been to hospitals
for paralysis and had tried galvanism and other

CLASSIFICATION OF BLADDER DISEASES IN MARRIED, SINGLE, AND STERILE WOMEN.

CONTENTS

OF

"URINARY DISEASES," by DAVID JONES, M.D.

(SECOND EDITION).

With Introduction by the Editor, GORDON GRIFFITHS JONES, Surgeon to the Home Hospital, 10, Dean Street, Soho, London.—Names and addresses of patients cured, see Appendix to the present Edition.

STONE, ENLARGED PROSTATE, STRICTURE AND OBSCURE BLADDER DISEASES IN MEN.

INCONTINENCE OF URINE.

IMPROVED TREATMENT.

lii

PAGE.

PART I.

DISEASES

OF THE

BLADDER AND PROSTATE.

GENERAL CONSIDERATIONS :—*The Bladder and the Diseases to which it is liable.*

THE bladder is subject to a variety of diseases. Authors classify them as " Acute " and " Chronic "; and these are, for the most part, curable and incurable. It is not the intention of the writer to incite an elaborate treatise on the bladder, nor to take cognizance of simple affections, which ordinary treatment will cure; but rather to draw attention to the *chronic* and as generally regarded *incurable* form of the disease, which has baffled the professional mind, and is looked upon as an opprobrium to the medical art. This disease, well known to the Profession and to the public, is, in women, called " chronic inflammation of the bladder or neck of the bladder," "irritable bladder," "catarrh of the bladder," " ulceration of the mucous membrane of the bladder," "nervous disease of the bladder," &c.

A

by overworked, and in their turn become diseased by diuretic medicines and stimulants.* The writer seldom sees a patient who does not say, "my doctor tells me to drink gin." One doctor advises "whisky;" another orders the patient "abroad, to drink the waters," these waters being usually irritating alkaline diuretics, according to the locality selected. It would appear most unreasonable to whip and spur a horse already jaded from overwork: common sense would dictate rest, which always does good; but, as your bladder is weak, the doctor whips up the kidneys by drugs, and the alcoholic drinks recommended, to secrete (manufacture) more urine; and thereby the poor crippled bladder (which is incapable of disposing of the normal quantity) is actually made to work more in a diseased and enfeebled condition than in a sound, strong, and healthy state of the organ.

The causes of these terrible diseases are numerous, and when weakness exists these causes may actively operate with serious effect. Among the superinducing or exciting causes alluded to, the following may be mentioned, viz., wet feet, lying or sitting on damp ground (particularly in females during certain periods), sitting for a lengthened time in damp clothes, suppressed skin action, as by sudden exposure from heated and crowded apartments to the cold air, sleeping in sheets not properly aired, indiscretion in early life, gonorrhœa, neglecting to empty the bladder through modesty or otherwise, using strong or improper injections (particularly in the male sex), drinking large quantities of wine or beer, taking diuretic medicines for a lengthened period, &c, &c.

When the bladder becomes *chronically* diseased, in either sex, it is one of the most distressing cases a physician can have to deal with, (1) on account of the indescribable discomfort experienced by the patient, and (2) because of the physician's inability (by the usual methods) to afford relief. Chronic disease of the bladder, as of other organs, more usually follows a previous acute attack,

---

* The medicines usually prescribed are, Buchu, Uva Ursi, Pareira Brava, Triticum Repens, Copaiba, &c.

from whatever cause, and is accelerated, of course, by any neglect or unsuccessful treatment ; especially where the more usually adopted medicinal remedies are applied—remedies which reach the diseased organ only after they have lost their virtue and efficacy. The familiar illustration of food introduced into the stomach may here be mentioned. It is well known that food undergoes, in the stomach, a variety of chemical changes. It is conveyed thence to the first portion of the small intestine, and is there acted upon by biliary matter and pancreatic juice. From the small intestine, the *chyle* (the name given by physiologists to the food in this stage, and so called from its resemblance to milk), is absorbed by the lacteals through the *mesenteric glands*. During its passage through these glands another change takes place, and ultimately it mixes with the venous blood and is conveyed to the right side of the heart, thence into the lungs, where it undergoes another chemical change. The blood is then conveyed to the *left* side of the heart, altered from a blue (venous) and impure kind, into a bright red (arterial) and pure kind, and thence, by the general circulation, to the kidneys and all parts of the body. So, likewise, medicine (taken into the stomach) following the course of the food, reaches the kidneys, where, with the food, it undergoes further change, and combines with the urinary secretion. Thus deteriorated (as to its medicinal properties) by its circuitous course, the medicine reaches its destination, but only to be speedily expelled as urinary secretion unfit for any other purpose in the economy. The only possible exception to this is where the medicinal remedy is supposed to act, and acts, powerfully as a specific. Clearly, the treatment most likely to be successful is the one which enables the physician or surgeon to reach the organ, and to attack its diseased part, by direct means,—as is done by the writer's method of treatment of cases of diseased bladder.

For upwards of thirty years the writer has drawn attention, by public lectures and otherwise, to the cure of consumption, asthma, and bronchitis, by inhalation,—in other words, communicating the healing agent direct to the lung structure itself, by the process of breathing * His successful application of this in practice was so great as to suggest very forcibly that the *supposed incurability* of bladder diseases might, not unreasonably, be attributable to what may be termed the *indirect* attempt at cure, that is, in administering medicines which reach the diseased organ only by the circuitous route above mentioned. Hence, in his own practice, he applied another mode of treatment, and finally invented a very simple apparatus, by which, without pain, and with but slight inconvenience to the patient, a medicated spray may be showered (so to speak) upon the very spot which is diseased. To this method of direct application of various suitable medicaments to the inner lining or surface of the viscus itself, he attributes the great success which has attended his treatment of such apparently hopeless cases as were many of those detailed in the present edition.

* See published Cases.

PART II.

ANATOMY

BLADDER AND PROSTATE GLAND.

To enable the reader to understand the writer's local treatment of the Urinary Bladder and Prostate Gland, a brief account of the anatomy and relative position of the organs concerned is here introduced.

I.—THE BLADDER.

The Urinary Bladder is a reservoir for the urinary secretion, which enters it, drop by drop, from the kidneys, through the ureters, one on each side. The Bladder is situated, when empty, within the pelvis. In the male, the neck of this viscus is surrounded by the Prostate Gland (hereafter to be described); behind the Bladder is the rectum. In the female the uterus is situated between the Bladder in front and the rectum behind. The Bladder is a musculo-membranous bag, and is formed of three coats — a peritoneal or external, a muscular or middle, and a mucous or internal coat. The peritoneal covering forms the false ligaments of the Bladder, and assists in keeping the organ in *situ*. This covering is reflected over the Bladder in such a manner as to leave a considerable space, in front and below, *uncovered*. When inflammation of this covering, or membrane (as it is called) occurs, it spreads very rapidly and dangerously. A wound penetrating the peritoneum produces acute inflammation of it. Nature seems to have anticipated that the Bladder might require to be punctured when the urinary fluid could not escape through the natural outlet, and has wisely left two spaces, one anteriorly, and the other inferiorly, *uncovered* by the peritoneum. The surgeon takes advantage of his knowledge of this fact, and occasionally punctures the Bladder in one or other of these spaces for the relief of the patient, as becomes necessary in Stricture of the Urethra, thereby avoiding the necessity and danger consequent on injuring this sensitive peritoneal membrane. The muscular or middle coat is formed of pale involuntary muscular fibres, and is much thicker than the (external) peritoneal or (internal) mucous coat. The muscular coat forms a thicker portion of muscular arrangement on the external part, called *detrusor urinae* (*detrudo*—to thrust out) which is said to expel the urine. There is also a marked band of muscular fibres around the neck of the Bladder, forming a sphincter, and called *sphincter vesicae*. The Bladder in a normal condition is capable of holding without inconvenience about a pint of urinary fluid. In obstructive disease of the organ, as Stricture, Stone, and Prostatic enlargement, the muscular coat becomes immensely hypertrophied (thickened). Cases are recorded where the muscular coat has become an inch in thickness. The muscular coat of the Bladder, like the muscular structure of the heart or any other part of the body, increases in proportion to the exertion the muscular fibres are called upon to perform. Muscular fibres are added to them, just as happens in the muscles of the arm of the blacksmith who uses his heavy hammer, or in the extremities of an opera dancer, whose *gastrocnemii* (muscles of the calves of the leg) are seen to be so fully developed through the exertion of dancing. This muscular coat (as, indeed, the mucous and peritoneal coats also) is capable of considerable distension. The Bladder becomes distended by its contents occasionally so as to reach as high as, and to be distinctly felt at, the umbilicus (navel). One case related in this edition (Case of H.H.) is a very striking one. It came under the writer's notice in 1873. He drew by catheter one hundred and twelve ounces of urine (5 pints and 12 ounces). There are cases on record where the human Bladder has been distended to an even greater extent.

The cause of this accumulation was due (as will be seen by referring to the Case) to retroversion of the womb, which pressed on the neck of the Bladder. In this case the whole of the coats of the Bladder were uniformly distended. The viscus occasionally becomes distended in an irregular manner into pouches or bags, as happens in obstructive disease of the Urethra through Stricture or enlarged Prostate. This condition is called "Sacculated Bladder." The mucous membrane insinuates itself between the layers of the muscular coat, forming two or three large Sacculi, or a great number of small ones. The noted CIVIALE of Paris, had a bladder in his possession covered with these, so as to resemble a bunch of grapes. PLANTER saw one bearing thirty-nine sacs—each one containing a calculus. They generally contain urine or muco-pus, and occasionally give rise to serious inconvenience, because the contents cannot always be emptied by catheter owing to the orifices of the sacculi being closed by bands and valve-like formations. The mucous membrane frequently becomes disorganized, giving rise to symptoms of retention accompanied by considerable constitutional disturbance. The fluid moreover infringes upon adjoining viscera occasioning serious discomfort, especially so on the rectum which is immediately behind the Bladder.

The mucous membrane, or inner coat (third coat) lines the whole of the interior of the Bladder. In health the inner membrane is soft and smooth, and of a pale rose colour—it is studded with minute follicles, most numerous near the neck of the Bladder, the whole surface being covered by epithelium. When the bladder is empty it is thrown into internal folds, or wrinkles—when distended, this arrangement accommodates itself to a larger quantity of urinary fluid. The mucous membrane extends upwards from the bladder into the ureters (two small tubes conveying the urine from the kidneys into the bladder). It is on account of this continuity of mucous membrane that it not unfrequently happens that patients suffering from urethral, prostatic, and bladder affections die from kidney disease. The writer has frequently traced "Bright's disease of the kidneys" to this cause, and thinks it very probable that the yellow granular degeneration seen after death in kidney disease, results from chronic stricture of the urethra, and kindred specific affections of the prostate gland and bladder. It must not be supposed that because stricture, and other diseases do not co-exist with "Bright's disease," that these diseases were not the *cause* of the kidney mischief. Pulmonary consumption, for instance, often begins in the form of nasal catarrh, but by the continuity of the mucous membrane it travels into the throat, from the throat into the larynx, and then into the lung structure, and rarely ceases till it destroys life. The disease, so to speak, creeps down insidiously from the nose into the lung tissue. What occurs in the nasal, laryngeal, and pulmonary tract of mucous membrane, happens also in the genito-urinary tract. A gonorrhœa (acute inflammation of the urethral canal) leaves a slight gleet (chronic inflammation of the mucous membrane of the urethra). This gives no inconvenience, probably, for twenty years or more, but it gradually results in stricture, or implicates at once the prostatic portion of the urethra, occasioning inflammation of the prostate (prostatitis). This gradually extends into the bladder: from the bladder it travels up the ureters into the kidneys, constituting "Bright's disease." Outwardly, the mucous membrane extends from the neck of the bladder, lining the canal over the prostatic portion of the urethra and from thence throughout the whole length of the canal to the meatus (entrance of the urethra), where it becomes continuous with the skin. The mucous membrane all over the interior of the body, whether it be in the mouth, the stomach, the lungs, or other parts, serves as a protection to the deeper parts, as the skin on the exterior of the body, serves as a protection to the sensitive true skin below.

There is another arrangement of the mucous membrane which is smoother, void of *rugæ*, and far more sensitive and vascular, called "*trigone vesicale*"—a triangular space close to the neck of the bladder, and situated on the most dependent part. If there be a stone in the bladder it gravitates on to this sensitive space, so that when the bladder is empty, the foreign body occasions considerable inconvenience, until the urine intervening drop by drop, between the mucous membrane and the stone serves as a temporary buffer. The pain on this account becomes less as the urine is secreted, until the next act of urination removes the intervening fluid buffer, when the stone again worries the sensitive "*trigone*" by coming into more immediate contact with it. The ease which patients afflicted with stone experience in the recumbent posture, in bed or on a sofa, compared with being in an erect posture, is easily explained on the same principle. The foreign body when the patient is standing, or walking or riding, falls by its own weight on this sensitive spot, while in the recumbent posture it rolls away from this sensitive "*trigone*" into the back part of the bladder, where the mucous membrane is less sensitive, and consequently a patient suffering from stone is easy at night, while a patient having prostatic disease is always worse in bed. The reader can readily understand that, as this spot has no *rugæ* it would feel the effects of distension more than the rugæated part—which (so to speak) need only be unfolded. The writer is inclined to think that distension of this sensitive smooth unrugæated spot gives rise to the natural desire to urinate. When a person does not respond to the dictates of nature in a reasonable time, this sensitive spot in the male is so stretched and disturbed, that it rarely recovers itself, and leaves a life long recollection of disobeying the laws of nature.*

The bladder is largely supplied by blood vessels, lymphatics and nerves which also supply the rectum, uterus and ovaries. This accounts in a great measure, for a disease in the rectum, uterus, or ovaries producing symptoms as of a bladder disease in the female, and of the prostate in the male.

Some persons with disease of the prostate, are more troubled with discomfort in the rectum than in the bladder. One patient consulted the writer some years ago for supposed disease of the prostate, and had been treated for such. His symptoms were essentially urinary, and with the exception of hæmorrhoidal discomfort, attended with frequent attacks of bleeding from the lower bowel, no one would have suspected that his urinary discomfort arose from the hæmorrhoidal mischief. On examination, it was found that the hæmorrhoidal veins were very large and congested and as the treatment did not satisfactorily remove the bladder discomfort, the writer suggested a removal of the piles which was acceded to, and very shortly the urinary symptoms subsided.

Many similar instances have fallen under the writer's notice. One patient who was supposed to have a bladder disease (but had no bladder disease at all) was cured by the writer, of the supposed bladder inconvenience, by curing an ulcer in the rectum, which gave little or no inconvenience to the bowel. One lady who had miscarried a great many times, and had been unsuccessfully treated for uterine mischief, was soon cured of the tendency to miscarry by curing her of disease of the lower bowel. This is readily accounted for. The same nerves that supply the rectum, supply also the bladder, uterus, and ovaries, and are derived from a plexus of nerves called "hypogastric." When the rectum is diseased there is a sympathetic connection between the rectal branch of the hypogastric and the vesical, uterine, or ovarian branches of the same plexus, as has already been mentioned.

---

* The writer has cured several of these cases by the Spray treatment.

Disease of the bladder, uterus, or ovaries in like manner produces an *apparent* disease of other organs through sympathy, and without great care in diagnosing each case the effect may be taken for the cause and the patient treated for a wrong disease.

II.—THE PROSTATE, OR PROSTATE GLAND.

This is a firm, glandular and muscular body, peculiar to the male. It is situated in front of the neck of the bladder, and derives its name from a Greek word—προστάτης (that which stands before). It surrounds the neck of the bladder, so that the commencement of the urethra (proceeding from the bladder) passes through it. It is placed, therefore, deep in the cavity of the pelvis. It lies below the bones of the symphysis pubis, and behind a ligament called "triangular ligament," a ligament which fills up a triangular space formed by the two opposite bones of the pubes, and from which it derives its name. The prostate gland rests beneath on the middle portion of the rectum—whence it is, that the surgeon can explore the dimensious of this glandular and muscular body by introducing the finger into the rectum (lower bowel) The prostate is formed of a right and left lobe, and a median portion. In shape and size the organ, in a normal condition, resembles a truncated cone, compressed from above downwards, and on that account it has been compared to an Italian chestnut, or ace of hearts—its base being turned backwards to the bladder, and the blunt apex forwards to the triangular ligament and that portion of the urethra called "membranous," which, joins the prostatic portion. The base of the prostate, its thickest part, is slightly notched in the middle and receives the common ejaculatory ducts from the testes and seminal vesicles. The size of the prostate varies considerably according to the age of the person. In early life it can hardly be discovered, and weighs only a few grains. As puberty approaches, and the organ is called into activity, it becomes larger, and in an ordinary adult it weighs from half-an-ounce to an ounce. After middle age, and in old age, it enlarges considerably, and in a diseased condition it assumes still larger dimensions—from the size of an orange to that of a cocoa nut. As the bones of the pelvis are in front of this organ, it cannot very well enlarge forwards, consequently it enlarges towards the bladder and rectum-structures which are soft and yielding, and offer no resistance—whence it is that the abnormal size which it reaches occasions such discomfort in both these localities. The Prostate is surrounded by a strong and unyielding fascia—pelvic fascia. When there is congestion or inflammation in this organ, or when, as sometimes happens, an abscess forms as the result of inflammation, specific or otherwise, it occasions one of the most painful diseases and gives far more discomfort than inflammation or abscess in other tissues not surrounded by bone and dense fascia. An abscess in a muscular structure, for example, has nothing to prevent its forming and pointing, consequently it soon gathers and breaks, which cannot happen in the organ under consideration.

The prostatic portion of the urethra is lined with mucous membrane continuous with the bladder behind, and the urethra in front. When the prostate gland is minutely examined it is found to consist (according to ADAMS) of muscular and glandular tissue. This muscular tissue forms an external layer below the fibrous capsule, and extends everywhere through the glandular substance —there is also a strong layer of circular fibres, continuous posteriorly with the vesical sphincter (a sphincter at the neck of the bladder which commands urination (and in front with a thin layer surrounding the membranous part of the urethra. The prostate gland consists of numerous small terminal follicles which unite into about twelve to twenty excretory ducts, which open by as many orifices on the floor of the prostatic portion of the urethra. This portion of the urethra extends towards the bladder and ends in a slightly rounded prominence, the uvula vesicæ, seen on the floor of the neck of the bladder. The reader will now understand that the prostate gland is in close contact with the Bladder, and

consequently, when it enlarges it does so into its cavity and interferes with the function of urination—it surrounds that part of the
urethra called (on that account) " prostatic urethra " which varies
in length from a little more than an inch to an inch and a quarter,
and extends as far as the membranous portion of the urethal
canal—a frequent seat of stricture. Besides the excretory prostatic
ducts that open into the prostate, there are seen at the base the
openings of the ejaculatory ducts through which the seminal fluid
is conveyed, to be further propelled by the functions attributed to
this glandular and muscular organ during the act of coition.

PART III.

———

STONE,

Classification, Description, Symptoms and Diagnosis.

—— ——

Stone in the Bladder the cause of great inconvenience and suffering—
may be present though unsuspected—originates in Blood Disease
—closely allied to Rheumatism and Gout—Deposits of Crystals
in Kidneys and Bladder—Hygienic principles applicable as
auxiliary means of Cure in early stages—Dietetic regime of
great importance—Food—Drink—Air—Exercise—Gall Stones
similarly induced—to be similarly dealt with—Direct means
desirable—for example, Inhalations of Oxygen, etc., etc., an
important element in the Treatment.

Stone in the bladder is specially recognized as the cause of the
most painful suffering. Some calculi, owing to their rough and
irregular surfaces, cause more suffering than others. Oxalate of
lime or mulberry calculus—so called from its being tuberculated
like a mulberry—produces more inconvenience than other calculi,
which are smooth though larger. In many notable instances,
however, very large calculi have been known to reside in the
bladder for a lifetime, without serious inconvenience. Mr. Cadge,

of Norwich, made a *post-mortem* examination on an old man who died of abdominal aortic aneurism, and discovered, accidentally, that the bladder contained a large stone, composed of *lithic acid*, weighing nearly nine ounces, and, still more remarkable, " the bladder was found healthy, and the mucus membrane pale." This patient had been attended by Mr. Cadge on and off for years, but so little trouble did the stone cause that neither doctor or patient was aware of its presence."*

Two instances are within the writer's own experience. One is E.B.'s case (see "Cases"). At first stone in the bladder was suspected, but the patient's recovery suggested doubt as to the accuracy of the diagnosis of the case, it being thought, naturally enough, that he could not have recovered had a stone been imprisoned in his bladder. Contrary to the writer's otherwise invariable practice, wherever any doubt exists, the patient in this case was not sounded for stone, his urethral canal being so unnaturally small as to refuse admission to the smallest sound. The applications used in his case were administered through the smallest French catheter. Happening to be in town some considerable time after his recovery he called upon the writer, and said there was nothing the matter with him but that he occasionally felt something roll about in the region of the bladder, without pain. He consented to the introduction of one of Otis's smallest silver instruments, and, after a good deal of difficulty in passing it, a large calculus was discovered as unmistakably present. This gentlemen, however, still believed otherwise ; and in reply to a letter urging him to come to town to have the stone removed, he wrote (under date 16th August, 1889) : I cannot make up my mind that I have stone : it seems to me that if I had I should necessarily be more plagued than I am." The writer is, nevertheless, *confident* that a stone existed in that case, though the spray treatment adopted so lessened the irritability of the mucus membrane of the bladder that the

---

* See *Lancet*, April 5th, 1878, page 472.

B

calculus resided harmlessly therein without injuring the viscus. Professional gentlemen reading this will perhaps, doubt the fact, and not unreasonably. The writer's extended experience in this speciality alone enables him to accept the conclusion as accurate—otherwise to himself also it might have appeared to be as unreasonable to expect subsidence of the symptoms of cystitis (inflammation of the bladder) while a stone existed in the bladder, as of inflammation of the eye containing foreign matter. The truth is, that although the symptoms of stone are generally very well marked, there are, nevertheless, many cases (and many which the writer has had to treat) where stone had not been suspected, and consequently, not discovered, until the process of "sounding" had revealed it. Many patients pronounced "incurable" have, in fact, simply been treated as suffering from " nervous irritability of the bladder, diabetes — Bright's disease, uterine disease," etc., etc. Very recently a patient came to the writer in great distress who had been indoor patient for many weeks in St. Bartholomew's Hospital, and in the Hospital for Women, Soho Square, and treated in both institutions for disease of the womb— the cause of disturbance and distress being in fact, a stone in the bladder, weighing two ounces and three-quarters, which the writer removed in the presence of several medical gentlemen in one hour and thirty five minutes.* The late Robert Liston used to say that whatever was the matter with the bladder it should always be carefully "sounded." There are many cases of stone in the bladder, and especially cases attended with hæmaturia (bleeding from the bladder), where the stone is hidden by clotted blood— villous tumour of the bladder—enlarged prostate, etc., etc. — cases in which the stone has not been discovered till after death. John Hunter discovered 20 stones in one bladder not discovered during the patient's life time ! Patients might, undoubtedly, in many cases, be spared an untimely death, while in others (as in

---

*Case of E. H. R. (see " Cases.")

a case reported by Sir Henry Thompson),† though suffering from incurable malignant disease of the bladder, the sufferings would be mitigated, by the removal of a stone.

In the other case alluded to (Case of I. B.—see "Cases") the patient was virtually cured of all his distressing *symptoms* of stone, and continued well for a considerable time, notwithstanding a stone was present in his bladder, and until the writer discovered the foreign body, and provoked inflammation and discomfort in the bladder, in the process of getting the exact dimensions of the stone, the patient was well satisfied with the cure. Since then the patient has had the stone removed, thus proving the writer's diagnosis of his case to have been correct. Unquestionably the treatment in both the cases referred to produced a healthy condition of the mucous membrane, and rendered it tolerant of the calculus, just as the *constitutional* peculiarities in the case related by Mr. Cadge permitted, for a life time, the residence in the bladder of a large stone, weighing nearly nine ounces !

Another patient (the late Mr. Salmon, of Tooting), was repeatedly told by Mr. Yeldham, Surgeon to the Homœopathic Hospital, Gt. Ormond Street, whom he had consulted, that in his case no stone existed. The writer, after only a cursory examination, felt assured that a large one was present. But the surgeon's adverse opinion prevailed, with the additional advice to "have nothing to do with Jones." Consulting the writer again, however, after years of intense suffering, the opinion was repeated, and the process of "sounding" recommended, but the patient recoiled from it. Soon afterwards, while the writer was in Paris, the patient became suddenly worse, and Mr. Savory, surgeon of St. Bartholomew's Hospital was consulted. The patient was sounded and the writer's opinion verified, but the patient was advised that at his age (80), if either "cut" or "crushed" he would succumb. Subsequently, getting still

†See Practical Lithotomy and Lithotrity, by Sir Henry Thompson, page 198.

worse, he was operated upon by another surgeon (Mr. Walter Coulson, surgeon to St. Peter's Hospital), and died under his treatment. The writer ventures to affirm that if this patient had been left under his care, and unbiassed, his life might have been spared, his stone being removable by a method of treatment free from the dangers attendant upon the ordinary methods. Others more seriously afflicted and older than the late Mr. Salmon was, have been similarly dealt with and cured (one still living, 86 years old).* Moreover, in cases of stone in the bladder amenable neither to lithotomy (the cutting operation) or to lithotrity (the crushing operation), the treatment adopted by the writer has been found useful, not only in prolonging life, but in greatly ameliorating the suffering even in the very worst cases.

Thus much as to the oft-times actual presence of stone in the bladder, though unsuspected. In the majority of cases, however, as soon as stone is formed in the bladder it worries the patient greatly in its repeated attempts to gain exit at each act of urination. The pain occasioned soon draws attention to its presence. When the bladder contracts during the expulsion of the last drops of the urine it grasps the stone, which, by its own weight, falls on the sensitive triangular spot of the viscus, called "trigone vesicale."

Stone in the Urinary Bladder, especially *lithic acid* and *oxalate of lime*, like stone in the gall bladder, originates unquestionably in blood disease. It is, in truth, an effect rather than a cause. The history of its formation may be stated thus :— The kidneys are the elaborators of urine, which urine they secrete from the blood. The blood is loaded with *lithic acid* or *oxalate of lime*, as the case may be, which nature is trying to depurate as excretory matter through the kidneys. In other words, stone could not be formed unless the various functions

---

*See cases 9 and 10 in the Author's 2nd edition of Urinary Diseases, to be obtained of the secretary, Home Hospital for Stone, &c., 10, Dean Street, Soho, London.

in the body failed to get rid of this morbid product by the process of combustion, respiration, and transpiration. It is notorious that " high " livers, and persons of sedentary habits, are more prone to calculous diseases than those who live moderately and take much exercise.

Rheumatic and gouty persons are within the same category. Gout and rheumatism are, indeed, also blood diseases, the blood being loaded with *lithic acid* and other morbid matters which nature fails to get rid of. Stone, moreover, frequently exists in association with rheumatism and gout. In some instances the rheumatism or gout may predominate, while in others the calculous tendency may be most apparent. Gout, rheumatism, and calculous diseases are really so allied that it is just an accident which of these diseases may most assert its power. One developes itself because there is a particular weakness and tendency to gout or rheumatism ; or stone in the bladder becomes the prevailing disease because of the preponderance of weakness in the genito urinary tract. But so closely are they allied that the one often co-exists with the other. While the urinary fluid is secreted in the kidneys it becomes concentrated, and one or more crystals are formed—layer after layer becomes deposited on the fragment which increases in size—the crystal soon becomes too large for the space in which it is located, hence it becomes dislodged, and seeks a larger residence in the pelvis of the kidney, and eventually is deposited in the mouth of the ureter (the tube which conveys the urine drop by drop into the bladder) and from thence, in favourable cases it is conveyed into the bladder. From the bladder it, in like manner (in favourable cases), passes away with the urine without further inconvenience, if, however, instead of being carried away it becomes imprisoned in the bladder, it forms a nucleus for additional layers of morbid product from the urine to be deposited upon, and in this manner it increases in size, occasioning more and more discomfort.

It not unfrequently happens that some patients pass large quantities of gravel, in the form of cayenne-pepper-like crystals, or "brick-dust" sediment, and, it may be, for many years without being much inconvenienced by it. These crystalline gravelly substances, or amorphous "brick-dust," sediment, deposit themselves in the bottom of the chamber utensil. Other forms of gravel, varying in colour and chemical constitution, frequently present themselves, which are typical of different calculi, such as phosphates, and their various compounds of ammonia, soda, magnesia, lime, mixed phosphates (fusible calculus).

The rarer forms of calculi need hardly be mentioned, which are associated with deposits of silicious matter—cystic oxide, xanthic, oxide, &c. Fibrinous calculus, described by the late Dr. Prout—uro-stealith calculus, described by Heller, in 1884, and Dr. Moore, of Dublin, as well as blood calculus formed of blood corpsules and phosphate of lime, found in the kidneys of consumptive people, are so rare as not to require comment in a work intended as much for popular as for professional perusal.

Every now and then, instead of passing into the bladder in the form of cayenne-pepper-like crystals, a fragment or crystal becomes *lodged* in the kidney, and while there receives additional deposit from the urinary secretion, until it attains large dimensions. In favourable cases this is dislodged, and forces its way into the bladder, through the ureter. This occasions excruciating agony and is called "passing stone from the kidney." In some instances this occasions more acute physical suffering than stone in the bladder. When the stone is within reach, as it is when in the bladder, it can be got at and the patient's sufferings relieved; but when, as sometimes happens, it becomes so large as not to be able to pass from the kidney along the ureter, the difficulty in treating the case is increased. Until quite recently cases of this kind were looked upon as hopeless, but now the calculus is sometimes removed by operation. The writer sees no reason why this operation should not be performed more frequently than it is if the impacted stone is large, prominent, and easily cut down upon. But even in these extreme cases much might be done to alleviate the suffering and dislodge the imprisoned calculus from its position—in other words, so to reduce the size of the stone as to allow it gradually to glide along the ureter into the bladder. Of course, while out of reach nothing can be done with the impacted stone, save the operation suggested ; but once brought into the open field (the bladder), it can then be effectually dealt with. Such cases are, however, usually regarded as hopeless by the Profession.

Several of such cases have been treated by the writer. In this part, dealing with "general considerations" only, the following may sufficiently indicate the reasoning pursued :—The blood is the *cause* of the mischief: it is loaded with the elements of *lithic acid* ($C_5 H_4 N_4 O_3$), which compound is derived from what we eat and drink, and it is allowed to accumulate and concrete into crystals, because there is not a sufficiency of watery material to hold it in solution, or of oxygen in the body to burn it into soluble *débris*. It is inspissated urinary product. Abstaining from articles of food and drink which abound in hydrogen and carbon, such as spirits, wine, fat, sugar, &c., and which consume a large quantity of oxygen taken into the system, will, assuredly occasion a more perfect combustion of *lithic acid*, and prevent ready deposition on the surface of the calculus. Another view leads to similar conclusions :—Some physiologists and physicians say that persons suffering from calculus diseases should abstain from animal food which abounds in nitrogen. This would, no doubt, have a beneficial effect, because thereby the oxygen in the system would be better able to burn up the ternary compounds just spoken of, viz., spirit, fat, &c. Abstinence from the *ternary* compounds would, nevertheless, serve a better purpose, as patients advanced in life can more readily subsist on the quarternary (nitrogenous) compounds (O. H. N. C.), such as animal food, bread, eggs, milk, &c., than upon the ternary (o. h. c.) or vegetable elements, the sole object being to occasion perfect combustion, and thus get rid of the superabundant *lithic acid* which forms the calculus, and for the same reason which led physicians some years ago to administer lemon juice as a remedy for rheumatism, viz., on account of its being highly charged with oxygen, which helps in the process of combustion, and thus renders more soluble the *materies morbi* occasioning rheumatism—*lithic acid* ($C_5 H_4 N_4 O_3$). Dr. Carpenter, in his work on physiology (page 627), alluded to this circumstance. He says:—"Thus, then, we have seen that the kidneys serve as the special instruments for depurating the blood of those highly azotized compounds (nitrogenous, fleshy, and albuminous foods) which are formed in the system by the decomposition of the materials of the albuminous and gelatinous tissues, and also by the non-assimilated components of the food. We have seen also that they serve for the removal of certain compounds, of which carbon is a principal ingredient ; and these, although normally present in but small amount, may undergo a marked increase in disease, especially when the liver is insufficiently performing its function, or the respiratory process is obstructed." Whichever view is taken, the application of these dietetic principles will most assuredly aid in the object to be gained, it being well known that the liver as well as the lungs and skin get rid of a large quantity of carbon.

Drinking largely, and for a long time, water impregnated with compounds containing oxygen, in the form of lemon juice, or citric acid ($C_6 H_8 O_7$), or inhaling oxygen gas — often administered by the writer with most beneficial results—must, in the course of time, burn up the lithic acid, reduce the size of the stone, and allow it to descend into the bladder.

Experience has proved this to be true in respect of rheumatism, and gout affections ; and the writer firmly believes that if the principle of this remedy were more extensively and perseveringly applied in cases of impacted stone, it would prove as beneficial and successful in other hands as it has in his own. For the same morbid product which induces rheumatism induces stone in the kidneys also, viz., *lithic acid*—and the same law holds good respecting the action of water (if distilled) on a stone in the kidneys, as is taught us by the old adage, " the constant dripping of water wears away a stone." Observation, indeed, shows that water wears away stones far harder than a *lithic acid* calculus.

In several cases under the writer's treatment the stones were dislodged. In one the patient strictly followed, for nearly four years, the advice and instructions given, when, after repeated threatenings, the stone passed into the bladder. From that time the patient has had no further evidence of kidney trouble, and has actually recently made a wonderful recovery from an attack of hemiplegia (paralysis of one-half the body) though in his eightieth year. The other, a man forty-one years old, after observing treatment two years, was suddenly seized with violent pain in the left kidney, and along the course of the left ureter, and in about two days after, he passed, while urinating, a good sized stone from the bladder. In a third case the patient, during a period of three years passed as many as 748 stones—the size of the stones becoming smaller each year, and necessarily less painful to the patient till ultimately he was apparently quite cured. That the treatment in these cases really caused the dislodgment, cannot, perhaps, be positively asserted; but reason and common sense, aided by chemical and physiological knowledge, lead one to suppose that the method of treatment adopted must have exercised considerable influence on the hitherto incarcerated enemy. It may, moreover, be confidently asserted that the treatment applied, and the dietetic directions observed, contributed to the successful result. It is equally clear that total abstinence from the elements that go to form *lithic acid* must have at least lessened the formation of further deposition on the surface of the already impacted calculi. In one case the continued deposit, extending over a period of four years, must have added considerably to the size of the calculus, had not the means adopted diminished its size, and so allowed it to pass into the bladder. This was effected without pain, thereby doing away with the supposition that dilatation or ulceration of the ureter had occasioned its descent. The constant dripping of water had reduced the stone, until it finally dropped into the bladder by its own weight. As other calculi have not since been seen in the cases alluded to, it strengthens the opinion that the treatment adopted was curative and preventative.

The writer may, perhaps, be permitted to say in this place that if he should ever himself become the subject of the like disease, he would unhesitatingly adopt the like treatment, with full confidence in its success.

Another treatment for rheumatism is the alkaline, such as potass and soda. Patients get well under this treatment also, because alkalies dissolve *lithates* and *lithic acid.* Where patients dislike acids the opposite plan of treatment might be employed. Patients go to the German and other Spas for rheumatism, stone in the bladder, and allied diseases and gain benefit. They are required to avoid stimulants, fat, etc., compounds of carbon, oxygen, and hydrogen (the ternary compounds), and to drink the alkaline and saline waters containing potass, soda, etc., which prove beneficial on the principle above referred to. In some cases a very long-continued use of these remedies is needful to produce the requisite change. Hydropathic treatment also does good in these cases ; and for precisely the same reasons. Patients abstain from fat, sugar, spirit, wine, etc., etc., and thus the great *lithic acid* feeders are effectually burnt up. A large quantity of water being conjoined, the blood becomes purer, and the insoluble *lithic acid* is rendered soluble. Exercise also is enforced, by which oxygen in large quantities is taken into the blood. Moreover, the skin is excited to get rid of hydro-carbonaceous compounds—*lithic* and *lactic acids*—which pass as effete matter through the numerous miles of tubules in the skin, and the result is, that patients return to their more town-like homes very much better for the treatment.

But, whether hydropathy be resorted to, or the "Spas" treatment, a few weeks or even months will oft-times be found insufficient for the purpose intended. Before rheumatism, stone, or gout clearly develope themselves the morbid products accumulate little by little, year after year, for perhaps, ten, fifteen, or twenty years. It would, therefore, be unreasonable to expect a cure within a few weeks, by such means alone. To cases of " gall-stone " the writer has likewise applied the treatment above alluded to with perfect success. In such cases the patient's biliary secretion (gall or bile) becomes very thick, containing too little water ; in other words, it is inspissated, and ultimately concreted, and "gall-stones " are formed. Reason alone suggests that thinning the blood, by drinking large quantities of water or some other harmless fluid, must prove beneficial. This happened in the cases above referred to. Viewing stone in the urinary bladder in the same light as gall-stone in the gall-bladder, viz., as being originally a blood disease, the wrtter has, as another element in, or part of his method of treatment largely administered remedies by inhalation into the blood, as oxygen, etc., etc.,—to prevent a recurrence of both morbid secretions, and patients have derived considerable benefit from such method of treatment. For external diseases, including every form of skin affection, it has also been applied extensively and with marked success. All forms of skin diseases from a rash to pimples, boils, and carbuncles, evidence some blood impurity, and to all such the writer unhesitatingly recommends and applies medication in a direct manner into the blood. Are not zymotic diseases, such as small pox, scarlet fever, measles, and fevers generally, taken into the blood by inhalation ? Ague and fever poison, common in the tropics, are also taken into the blood by direct inhalation.

If, therefore, infinitesimal matter thus taken into the blood—Small Pox for instance—be so potent in inducing disease, why should not remedies for skin diseases, though administered infinitesimally by the process of inhalation be equally efficacious in curing disease. Whence comes the renovating power of the seaside or mountain top? It is the pure air highly charged with oxygen, iodine, bromine, etc.—these elements go into the blood and by direct contact purify it—through the very channel ordained, and through which alone the vital fluid can be properly purified—Away then, with your "blood mixtures," quinine, and tonics swallowed into the stomach, frequently never (as medicaments) reaching the blood at all.

Calculi—commonly called stones—have been classified as Primary and Secondary.—The primary are those which form in the kidneys, and are secreted direct from the blood, viz. (1) *lithic acid* which was discovered by SCHEELE in 1776 and is composed of carbon, hydrogen, nitrogen, and oxygen, in the following chemical proportions $C_5 H_4 N_4 O_3$. and (2) *oxalate of lime* discovered by WOLLASTON in 1797—a combination, as will be seen, of oxalic acid ($C_2 H_2 O_4$) and lime ($C_a O$) represented chemically as follows :—($C^a C_2 O_4$.) This, from its resemblance to a mulberry, has been called " mulberry " calculus. The secondary are *not* formed in the kidneys, but, as their name implies, secondarily in the cavity of the bladder—from the urine. A morbid state of the bladder, and its appendages, occasions retention of urine in the bladder, which urine becomes decomposed into ammonia ($N H_3$) and other compounds.

In the mucus of an inflamed bladder is found also phosphate of lime ($C^a_3 P_2 O_8$). This combines with phosphate of magnesia ($Mg_3 P_2 O_8$) a constant urinary product.

All these compounds chemically combine with each other—the result being the formation of a stone called "triple phosphate" or "ammonio magnesian phosphate" ($NH_4 Mg PO_4$). These secondary calculi frequently co-exist with prostatic disease. The prostate gland becoming enlarged forms a mechanical impediment to the flow of urine which, kept in the bladder, becomes stagnant and decomposes—and hence, as already described, secondary calculi form.

The calculi most frequently coming under the surgeon's notice are of three kinds (1) lithic (or uric) acid calculus ($C_5H_4N_4O_3$) forming about three-fifths of all the calculi (2). Phosphatic calculus—a combination of phosphoric acid (H_3PO_4) with the volatile alkalies and the alkaline earths, forming about two-fifths —and (3) oxalate of lime ($Ca\ C_2\ O_4$) or mulberry calculus forming about three per cent. only.

Some authors say that lithic acid calculi form fully ninety per cent. of all the calculi—and the writer's experience confirms this estimate as being much nearer the truth than the three-fifths given by more recent authors. Persons subject to this form of stone have usually good general health, look hale and hearty—"the picture of health"—the urine is free and abundant in character, and the individual comforts himself in this, and tells you with an air of satisfaction that he is "all right" and can "eat and sleep like a child." He tells you, nevertheless, that he has been passing gravel for months, in the shape of cayenne-pepper-like crystals. These form in small or large quantities, according to habit and circumstances, and, it may be have been passed without inconvenience. They are formed in the uriniferous tubes, at their orifices and around them, and in the calices of the kidneys. So long as these crystals continue small and pass with freedom, things go on seemingly right enough. By and by, however, the crystals become larger, aggregate in greater quantities, and infringe on the functions of the kidneys, thereby occasioning irritation and inflammation, followed by a sense of pain in the back—temporary inflammation is established, causing great discomfort, and until the crystals are dislodged, the patient suffers great inconvenience. Occasionally much larger crystals form which inflame and distend the mucous lining of the kidneys, and stretch the tubes of the organs as they endeavour to dislodge themselves. This occasions a flow of blood, at first in very small quantities, and recognizable only by microscopic examination. A larger fragment than usual not unfrequently calls the patient's attention to his condition—he is suddenly seized with violent cutting tearing pain in the urethra—the flow of urine is somewhat impeded—he passes blood in visible quantities, and then sends for the doctor who finds that a small irregular sharp-edged fragment of uric acid is lodged in the urethra. Sometimes the fragment is violently expelled during the act of urination, and, to the patient's delight, rattles in the chamber utensil. This, however, is only the commencement of his troubles. He must be told that he may have a similar impaction in the ureter, attended with far more inconvenience. Or, he may have a fragment in the structure of the kidney itself, which refuses to be dislodged and this is much more serious. If it become free and descend into the bladder, he will have much to be thankful for, since it is then within reach, and can be dealt with.

Stone in the bladder may occur at any period of life, from childhood to old age. It is very common among children of the poorer classes. Fully one-half of the cases admitted into Guy's Hospital are children from the very poorest districts, and consequently among the worst fed. It very rarely occurs among children of the well-to-do, but is very common among the middle and upper classes, between the ages of 50 and 75.

The symptoms of primary stone, or renal calculus during its journey from the kidney to the bladder are, as already stated, indicated by severe cutting pain in the loins and along the ureter, attended with considerable constitutional fever. When a rough stone, such, for instance, as a mulberry calculus, descends, it passes with considerable difficulty. After some nephritic irritation the patient is oftentimes suddenly seized with excruciating agony in the loins, along the groin, and to the testicle of the corresponding side, which is often retracted. This is accounted for by sympathetic irritation of a small nerve in front of the spermatic cord—

the genital branch of the genito crural nerve. There is also, sympathetic pain down the thigh. The writer has seen patients roll on the floor in agony, cold sweat meanwhile pouring down their faces. In the same circumstances the patient may also vomit violently, through nervous sympathy with the pneumogastric nerve. There is bloody urine from irritation of the bladder, constant micturition, and sympathetic pain at the end of the penis for about an inch from the entrance of the urethral canal, also in the lower boundary of the abdomen and pelvis. This pain is frequently present in enlargement of the prostate, impacted stone in the kidney, chronic cystitis, cancer and villous tumour of the bladder, and indeed in almost every form of bladder and prostatic mischief.

When the stone is loose in the bladder the symptoms are not so acute. The stone rolls about and settles in the most dependent part of the bladder, behind the prostate, producing mechanical irritation of a part called *trigone vesicale*, a triangular vascular and sensitive spot at the neck of the bladder ; the character of the pain is described as a dull wearing weight, extending upwards to the lower part of the abdomen and downwards along the perinæum. It occasionally extends down the thighs—and sometimes also shoots along the penis and fixes itself in the glans penis. Occasionally, too, there is remote pain in the knee, foot, heel, or even the arm. The patient feels worse when the bladder is empty, as the stone falls on the sensitive " trigone vesicale," and the bladder, in trying to expel the last drops of urine, contracts violently on the stone, causing greater pain. As the urine accumulates in the bladder the fluid intervenes between the sensitive spot already spoken of, and the calculus, and relieves the pressure. The discomfort is always greater during severe exercise or quick movement of the body, as the foreign body shifts up and down during any kind of movement, and causes pain. Although pain during exertion is very symptomatic of stone in the bladder, still the writer can recall two instances where the patients were always better when walking about, and even when riding in omnibuses. Strangely enough both these patients suffered more when in bed and resting in the recumbent posture. In cases of difficult diagnosis this peculiarity must not be overlooked. Both the practitioners who had been treating the patients in question had assured them that they were not suffering from stone. As the stone increases in size, or becomes roughened by deposition, more irritation is established, the urinary secretion becomes cloudy, and mucus is deposited in the chamber utensil as evidence of cystitis (inflammation of the bladder). In some cases the practitioner is not unfrequently misled in his diagnosis owing to the urine being clear. It is so sometimes, when the stone is very compact, smooth, and small. The patient consults the doctor, not because of any great inconvenience but as for a slight "malaise," and there being no cloudiness or blood (almost always present in stone) the doctor mistakes the case, and the stone remains undiscovered.

Some years ago the writer discovered, in a case of this kind, a small stone not larger than a horse-bean, which had, nevertheless, given great discomfort. The stone was at once removed, to the gratification of the patient, and annoyance of a local practitioner of considerable reputation. When there is any doubt about such a case a shrewd practitioner will never fail to institute a chemical and microscopical examination of the urine. This will most probably reveal blood globules and crystalline matter not visible to the naked eye. Should the above means not elicit a satisfactory diagnosis the practitioner proceeds to use the mechanical test, the "sound" as it is called, which is described by a French authority, as a long finger which probes about in various directions, till it discovers the calculus, if any really exists. It is a matter of great moment to discover a stone early, while it is (so to speak) in its infancy, for then it can be at once dealt with. If allowed to remain, it grows and becomes more stubborn—and, of course, much more difficult to manage. This growth, almost unsuspected, usually occurs in persons who are tolerant of pain, and who have the appearance of sound health—the surgeon being thus thrown off his guard. He cannot however be too careful in cautiously and *thoroughly* examining every patient who comes under his notice, whatever the nature of the case, or whatever his or her position in life may be. The writer well remembers a case in which he was materially guided to a successful treatment (giving him considerable reputation) by listening attentively to a long history given by the patient of her malady, which up to that period had been regarded as constipation and hœmorrhoids. She said (among other things) "I have a constant bearing down pain in the lower bowel, as if I had something as large as a child's head there." This naturally enough suggested an examination of the rectum, the result being, discovery of a large mass of feculent matter which must have been lodged there for a very considerable time. It was removed by mechanical means, though with unusual difficulty and under the influence of chloroform. Purgatives had been prescribed for some years by eminent physicians and surgeons, as well as by other eminent medical men, without any relief. No pains had been taken to examine the patient. If suffering from a sore throat she would no doubt have been politely requested to open her mouth, to "see what was the matter." An examination of the lower bowel was however a different affair, and so the poor woman suffered indescribable misery for years—misery, the *cause* of which might if ascertained, have been removed in fifteen minutes. Resuming however the description of stone symptoms, it may be remarked that blood appears from time to time in the urine, and deposits with the mucus which is streaked with it. Occasionally blood appears in very large quantities, greatly alarming the patient. A patient declaring that he had none of the ordinary symptoms of stone (that which attracted his attention being the passage of a large quantity of blood) was treated as for rupture of a blood vessel in the bladder. "Sounding," however, at once discovered the cause of the bleeding, which soon ceased when the cause was removed.

c

Patients suffering from stone are nearly always better in bed. The stone in these cases rolls from the sensitive neck to another portion of the bladder, and the discomfort ceases. As the patient gets into a standing posture the discomfort returns. In disease of the prostate, the patient is on the contrary always worse at night, and better in the day time. Another symptom of stone is this : while the patient is in the act of urinating, the fluid as it passes along the canal sucks the stone into the neck of the bladder and suddenly stops the flow of urine—the stone however soon rolls away by its own gravity, and the urine again flows without interruption.

Where some patients pass water while in a lying posture, the stone also rolls away from the direction of the current and the urine flows uninterruptedly. When stone occurs in children the symptoms are similar. Owing to the discomfort at the gland of the penis, children frequently get into the habit of constantly pulling the prepuce, whereby it becomes elongated and enlarged. In a case having none of the usual symptoms of stone, the writer was consulted for a habit the child had contracted from an early period of wetting the bed. Finding the usual remedies for such cases unavailing, and noticing the state of the prepuce, he "sounded" for stone which was readily found. The process of "sounding" is after all that which gives the most positive indications of the presence of stone. A "sound" is a polished steel instrument shaped like a catheter, but with a much shorter beak and more acutely curved. In is introduced into the bladder more delicately by an experienced hand, and may be moved about in the bladder in all directions with little inconvenience to the patient. It has a bulbous portion at the extremity, the stem being made thinner for facility of movement in the urethral canal. If a stone be present it is reached in the majority of cases, and a distinct click or tap may be heard by both patient and manipulator.

Of late years, electricity—by means of the cystoscope—has been somewhat extensively employed as an illuminating medium for examining the bladder ; but the limited field of vision it affords, renders its adoption as a diagnostic agent—except in some few obscure cases—of less value than was anticipated.

The old-fashioned " sound " was a very imperfect instrument, and the stone unless very large was not discovered. For the present instrument we are indebted to Baron HEURTELOUP. The bulb portion of the "sound " is thicker than the stem, and was introduced by the late Sir WILLIAM FERGUSSON. The greatest care and gentleness should be exercised in the use of this instrument, which should only be handled by an expert. When the patient is very sensitive to pain, it is better to administer ether and sound the patient most thoroughly while you are about it.

Although this physical test is as perfect as it can be when the stone is touched, there are notwithstanding instances where patients have given full evidence of calculus, yet when examined no stone has been discovered. There are other instances where able surgeons have discovered stone and have distinctly heard and felt it at the first sitting, but entirely failed to do so subsequently. The bladder sometimes acts towards a foreign body in the most eccentric manner—especially in sensitive persons. The writer, on one occasion demonstrated to a certainty the existence of stone, and clearly made out the dimensions and nature of it in the presence of others who afterwards verified it, yet subsequently failed to discover the same stone, feeling confident that it was in the bladder notwithstanding. The bladder has peculiar move-ments of its own when under examination by instruments, and seems to secrete its lodger in a most tantalizing manner, defying detection. Sir HENRY THOMPSON has appropriately called it "playing at hide and seek with you." In these circumstances the administration of ether arrests this peculiar behaviour of the viscus and facilitates the discovery.

Stone in the female is by no means so common as in the male. The urethra in the female being larger and shorter, calculi of large size have passed without much inconvenience. Occasionally, however, the stone does not pass. In most instances of stone in the female occurring in the writer's practice it has happened in women who have had large families. The womb has been more or less displaced and the anterior wall of the viscus has been dragged into the vagina, forming a cul-de-sac. This condition—"cystocele," as it is termed—has given ready lodgment to the calculus which, on that account has not presented itself at the neck of the bladder during the act of urination, and consequently has not been expelled as is most usually the case.

The womb in a normal position is situated immediately behind the bladder. When the bladder is distended, the womb which is in the median line alters the shape of the bladder, and projecting (so to speak) from behind leaves a cul-de-sac on each side of the median line. CIVIALE described this " bas fond " on either side of the central prominence. In women who have borne large families, and have gone a long time without emptying the bladder, the culs-de-sac on each side are considerably enlarged and dilated, consequently the bladder loses its contractile power. These two recesses behind, like that in front, frequently give rise to the lodgment of stone in the female bladder. One patient had been sounded by two eminent surgeons, both of whom said there was no stone. The writer examined the same patient and concurred in the opinion. The symptoms, however, pointed so strongly to calculus that, on a subsequent visit he gave the patient ether, and a small stone was readily found in the left sulcus. The mucus membrane of the bladder was in a most irritable condition and rebelled persistently when the " sound " was used. Ether overcame the sensitiveness of the bladder. This patient had considerable displacement of the womb which partly projected out of the vagina, serving as an impediment in discovering the stone. The truth is that in such cases the practitioner is easily misled. There is also another reason why stone in the female bladder is frequently overlooked Women are very subject to displacement of the womb on to the neck of the bladder giving rise to constant micturition, and not very subject to stone (for reasons already mentioned), and displacement of the womb is therefore the sole conclusion usually arrived at   This conclusion pronounced in so many cases coming under the writer's treatment, had been arrived at in two of the cases reported in this edition. In one the stone was discovered and removed, in the other the patient was perfectly cured of a malady to which she would have soon succumbed had the cause not been discovered and remedied.

PART IV.

SECTION I.

STONE,

Its Treatment and Cure.

THE history of the various means employed in the treatment and cure of stone in the bladder dates from a very early period. From a remote period indeed and even to the present time, attempts have been made to dissolve stone by secret and other remedies administered in large quantities through the medium of the stomach. Besides medicaments thus administered, the operation of cutting for stone was also practised, and with considerable success, notwithstanding it was performed in a very primitive manner and by uneducated persons.

We have an account of this operation as early as the Augustine era—it having then already been performed some hundreds of years among the Greeks and Romans. HIPPOCRATES (B.C. 460) alluded to it in his works. He compelled his pupils to take an oath to the effect that they would not practise lithotomy, and advised them to leave it to those who were "specially famed for it." In those days it was, as already mentioned, practised by uneducated persons, the operation being called "cutting on the gripe "—referring to the mode of operating. The stone was seized in a very rude manner from the back part of the bladder, through the lower bowel, roughly "griped" with the ends of the fingers, and reached by cutting from the perineum—the patient being firmly held by two men sitting side by side—hence the term "cutting on the gripe." In the course of time the practice fell into the hands of more competent, though still uneducated and as it were itinerant operators, who invented a variety of instruments for the purpose, and ultimately one of the fraternity Frère Jacques, studied anatomy in Paris and became an accomplished anatomist. But increasing acquaintance with anatomy, and perception of the intricate nature of the operation he had so often successfully performed while comparatively ignorant of the anatomy of the parts implicated, is said to have intimidated him, and his operations were ever after less successful. From the time of Pliny to the present day a variety of solvents for stone have been suggested, composed chiefly of alkalies from the animal and vegetable kingdoms which, though worthless as solvents for stone in the bladder, were nevertheless, as already mentioned (in Part I.) useful as preventing further enlargement of the stone by checking greater deposition of matter. Even at the present day the experiments of Sir Wm. Roberts of Manchester prove that small stones of uric acid may, under certain conditions, be somewhat reduced. The result however is on the whole so unsatisfactory in respect of positive cure that no enlightened practitioner ever thinks of attempting the experiment. Sir Henry Thompson in his clinical lectures on diseases of the urinary organs has given an interesting account of these so-called solvents, and comes to the conclusion that "if the stone be large the solution is impossible."

Until the year 1824, lithotomy, or the operation of cutting for stone in the bladder was the only plan of treatment adopted with a view to an effectual cure. In that year and in the presence of a committee of the Academy of Medicine, Civiale operated for the first time on two living patients by lithotrity, or the *crushing* operation. This method (lithotrity) has since CIVIALE's day undergone considerable improvements, both in respect of the instruments employed and the facility with which the operation is performed.

It is not necessary to detail the various changes and improvements that have taken place, or to describe the opposition with which the advocates of LITHOTRITY had to contend. It may suffice to say that the operation is now admitted by its strongest opponents to be most successful.

Thus the treatment for curing stone consisted, until very recently, of the two operations above named—LITHOTOMY and LITHOTRITY. Very large and very hard calculi are generally selected as suitable for the cutting operation (LITHOTOMY)—small calculi, up to about the size of a chestnut being treated by the crushing operation (LITHOTRITY). In calculi among young children, and up to the age of 20, LITHOTOMY, or the cutting operation has proved very successful, but not so in cases from the age of 50 and upwards. Age is, therefore, a very important matter to be considered when deciding which operation to select. Regard to the size of the stone to be got rid of is equally important, it having been well proved that when the stone is larger than about the size of a chestnut, and of the hard kind (Oxalate of Lime) LITHOTRITY is liable to become fatal in proportion to the size of the stone and the age of the patient. A further objection to LITHOTRITY is that a repetition of sittings is necessary, varying from five to ten, and even up to twenty. The cases are very rare where one sitting has sufficed —as for instance where the stone has been very small and soft, luckily seized by the lithotrite in a fortunate direction, and there and then thoroughly pulverized and the *débris* passed away without much inconvenience. The writer has not however met with such cases in his own practice.

The difficulty of LITHOTRITY has hitherto been the removal of the débris. Where the stone is large and hard, the fragments fly about the bladder in all directions. In the case of an oxalate of lime calculus the fragments are like bits of flint: they terribly cut and tear the sensitive mucous membrane of the bladder and urethra, and fatal consequences occasionally result. This and other circumstances led Professor BIGELOW, of Harvard University, U.S., to endeavour to improve upon LITHOTRITY, which in the writer's opinion, he has unquestionably done. His innovation upon LITHOTRITY (for the profession at present regard it as such) Professor BIGELOW calls "LITHOLA-PAXY," which signifies "Rapid Lithotrity with evacuation." He removes the majority of calculi by crushing with large instruments, in one sitting under the influence of ether. When however the stone is very large, or there is more than one stone, he does not hesitate to increase his sittings to two and three, or to prolong them to nearly four hours.

Moreover, his plan of treatment not having yet been tried on a large scale, the professor wisely guides himself by a most careful consideration of the tolerance (so to speak) of each particular patient. One case which he reports, in his Essay on the subject, is here subjoined :—

"Case 8 (Dr. C. B. Porter's case) August 12th, 1877. Aged 61.

" A large flabby man, with a feeble pulse.

Date of symptoms, twenty-six years.

Two stones, one of which is so large that it is barely possible to lock the lithotrite.

Passes water every fifteen or twenty minutes.

Three sittings. First sitting :—duration, one hour and a half under ether ; size of tube, twenty-eight ; quantity removed, two hundred and twenty-eight grains ; passed afterwards one hundred and eight grains. Second sitting ;—interval, four days ; duration, three hours under ether ; size of tube, thirty ; quantity removed, seven hundred and forty-four grains ; passed afterwards sixteen grains ; no after symptoms of importance. Third sitting :—

interval, five days; duration, three hours and three quarters under ether; size of tube, thirty-one; quantity removed, seven hundred and six grains; no pain nor discomfort afterwards; total number of grains, after drying, one thousand eight hundred and two. Result: discharged well, two weeks from the date of the first operation; after a few weeks the patient could retain his water from three to four hours."*

From the time of CIVIALE to the present day the teaching of Lithotritists has been to the effect that the bladder and urethra are very intolerant of instruments. Even Sir HENRY THOMPSON has spoken of "A sojourn, say of two minutes, in the bladder, which I will allow you, although you know I do not occupy so much time myself."†

The publication of Professor BIGELOW's views greatly delighted the writer, finding therefrom, as he did, that he also had been working in the same direction. For the last twenty-five years the writer has studied the question of dissolving stone in the human bladder. Its possible accomplishment has been universally laughed at and ridiculed. All the ordinary attempts for the purpose have consisted in the administration of alkalies and acids directed to the interior of the sensitive bladder, and the administration of the same remedies by the stomach. The commonest form of stone, lithic acid (said to be the basis of 19 out of 20 of the calculi), is so insoluble that the corrosive alkalies or acids necessary to dissolve it would burn through the tissue of the bladder and destroy life, long before producing the slightest effect upon the stone. Viewing the solution of calculus in this light it would of course be an impossibility. To solve the difficulty the writer worked in a totally different direction. His idea was to *isolate* the stone so as to get it under perfect control,

---

*See Bigelow's Essay on Litholapaxy (1878), page 29.

†See " Clinical Lectures on Diseases of the Urinary Organs, etc." by Sir Henry Thompson, 4th edition, page 188 —But for his present opinion see p. 45, also the " Lancet," January 17th, 1880, page 79.

and when thus isolated, or imprisoned, and away from the sensitive mucous membrane of the bladder, to act upon it with the requisite solvent. He calculated that a stone too large, or too hard for crushing—one for instance only suitable for LITHOTOMY —would require the residence of instruments in the bladder for one, two, or more hours, according to the size and solubility of the calculus.

Professional gentlemen reading this, will no doubt be inclined to put the same question as was put by M. MALLEZ when the writer mentioned the subject to him in Paris in 1878—" But, my friend (he asked), how are you going to get your platinum machine (the only metal that will resist corrosive acids and alkalies) into the bladder through the small urethral canal ?" The answer now, as then, may be—" that is my business "--the writer meaning thereby that in the present stage of his experiments, and under present circumstances (more particularly alluded to in another part of this work) he does not deem it desirable, or feel called upon, at once fully to reveal the *modus operandi* by which he proposes to apply his process. Suffice it to say here that for many years he has, in association with the successful treatment and cure of cases regarded as hopelessly incurable, demonstrated that the human bladder and urethra are not *so* sensitive as Lithotritists have supposed—his own method of treatment being nevertheless, such as to protect the mucous membrane of the bladder from any possible injury. When it is considered that the human bladder ofttimes retains calculi of enormous size, and for many years, it is absurd to suppose that beautifully polished instruments cannot be manipulated within its cavity for more than "two or three minutes" at one sitting. True, some patients faint when the smallest instrument is introduced for the first time. Some indeed faint at the mere sight of a polished steel surgical instrument, but in the majority of instances this sensitiveness soon subsides. Such instances moreover are only exceptional. Some people are so peculiarly constituted that they turn pale and sick if they smell a rose. It is not so however with the majority of mankind. Again, when a bladder has been accustomed to a stone for some time, it is often more tolerant of instruments. A bladder accustomed for years to a calculus of extravagant proportions, such as we find mentioned by some old authors, would certainly not rebel against the introduction of a fine polished instrument so readily as modern lithotritists would lead one to suppose. We hear of calculi so large as to nearly fill the whole cavity of the bladder. Enormous calculi have been found after death in the horse, and notwithstanding which, the animal lived a fair average equine life, and did as much work and as nimbly as horses without such an encumbrance. The writer firmly believes that the injury sometimes resulting from irregular and pointed fragments of stone remaining after the operation of LITHOTRITY has been performed, has been mistaken for effects resulting from the use of the instruments, and he has the greatest possible confidence that BIGELOW's method of curing stone will, if fairly put into practice, soon supersede LITHOTRITY as now ordinarily applied. The readiness with which the writer removed a large stone, weighing two ounces and three-quarters, in one hour and thirty-five minutes, and another very hard lithic acid calculus weighing very nearly one ounce, in less than an hour, and his success in removing a very hard oxalate of lime stone, weighing nearly four ounces (all which cases, with others, are reported in this present edition), gives him the greatest confidence in the future of LITHOLAPAXY.

Just as LITHOTRITY had to fight its way into professional favour amidst prejudice and jealousy combined with ignorance, so will it be with LITHOLAPAXY. Much however may be effected from the recognition of it already accorded by some leading lithotritists. Mr. THOS. SMITH, Surgeon to St. Bartholomew's Hospital has said :—" it is possible or rather, highly probable that it (BIGELOW's principle) will entirely reform our practice in the treatment of stone in adults "—and further, that " if BIGELOW's practice be founded on truer views of the whole subject (as it seems likely that it is) the domain of lithotrity has been greatly enlarged, and a real advance will have been made in the treatment of stone in the bladder."*

Sir HENRY THOMPSON speaking of BIGELOW's method says :— "This was a bold, but I believe it was also a happy idea. My mind was already prepared by past experience to receive it favorably, although the means BIGELOW employed in the shape of instruments, especially the lithotrites he proposed to use for the purpose, it was impossible for me to approve."†

In his last (5th) edition of "Clinical Lectures on Diseases of the Urinary Organs,"‡ page 175, he says :—" Having referred to BIGELOW's aspirator, I may briefly state that he has recently proposed to remove, at one sitting, all calculi of any size, and whatever may be the condition of the patient, by means of large lithotrites and the aspirator combined, devoting two hours or more, if necessary, to the purpose. Of this proposal I feel compelled to say that although the results may often be successful, it is to be feared that they must sometimes be disastrous ; for although there is no difficulty in achieving the object, as far as mechanical power is concerned, we cannot overlook the fact, that the vital conditions under which we are compelled to work must often limit the employment of mechanical force. I am free, however, to confess that the proposal to remove a large and hard stone at one sitting is an attractive one. I only fear whether we may not by adopting the system under consideration, pay too high a price for the purpose of attaining the end proposed. And in reference to this, I am bound to say that my own system has for a long time been gradually inclining to the practice of crushing more calculus at 'a sitting and removing more débris by the aspirator than I formerly did. Thus I have, for some time been in the habit of using in every case two lithotrites alternately in

---

*See " Lancet," January 10th, 1880, pp. 43 an l 45.
†See " Lancet," January 17th, 1880, p. 79.
‡ J. & A. Churchill, London, 1879.

the manner already described. With these light and handy instruments, which pass with the utmost facility, employed in this manner and followed by the aspirator, I can certainly remove calculous matter from the bladder more safely, and much more rapidly than with any large and unwieldly instruments."

The teaching of lithotritists has, as already mentioned, been to remove calculi not larger than a chestnut by crushing—the larger calculi being left for the cutting operation, which is *fatal* in about one in seven cases. Small calculi can usually be easily crushed. Sir Henry Thompson has done it successfully with his "light and handy instruments" — and encouraged prolonged sittings also. He says :—"The longest sitting I have as yet ventured on is twenty-five minutes, during which I removed 329 grains."* This is very different to his former teaching :—" a sojourn of two minutes."

With small calculi these instruments have ample mechanical power. Neither Bigelow or any other operator would need stronger power for such small sized stones. But both Bigelow and the writer have prepared instruments for dealing with such calculi as cannot, with safety, be crushed by "light and handy" instruments.

The two cases above referred to as treated by the writer,† where the calculi weighed two ounces and three-quarters in one case, and five grains under an ounce in the other, could not with safety have been crushed by smaller instruments than those suggested by Bigelow. We have heard of instruments breaking and bending in the bladder, although they were manufactured by good instrument makers. They were "light and handy" but the calculi were too large and hard for such light contrivances, consequently the patients had to be cut not only for the stone but also for the removal of the injured instrument, which further endangered the patient's life. The writer cannot see the force of the objection

---

* See " Lancet," January 17th, 1880, p. 79.
† See page 43.

raised against the larger lithotrites—on the contrary, he sees immense advantage in the use of them. In the first place, it crushes the stone most effectually, because it has ample mechanical power, and ought not to injure the bladder at all, if dexterously handled. The greater the power the greater the effect, and in this case the greater the safety in avoiding any injury to the instrument. In the second place, being strong, it enables the surgeon to work with confidence knowing that however large or hard the stone be, the instrument cannot come to grief. Take for instance a case, such as is related in Sir HENRY THOMPSON'S last edition, page 143,*—" I crushed the stone four times, bringing away a good deal of phosphatic material. I soon noticed that my lithotrite never went through the stone; it always went a certain way and then there was a hard mass. After four sittings I could not crush any more. It was clear that there was a very hard centre stone on which my strongest lithotrite had no impression, the crust only having been removed. I know from experience, the recoil of the lithotrite from an oxalate of lime stone so well, that I had no hesitation in saying such an one was present. Accordingly I performed lithotomy and removed a well-marked specimen of that kind." In another passage, he adds,—" An oxalate of lime stone communicates a sensation when grasped by the instrument as if you were laying hold of a piece of iron—you make little or no impression upon it." Now, one of BIGELOW'S instruments used in a case of this kind, would, the writer doubts not, have crushed the calculus by a few turns of the screw, and the patient would have been spared the operation of cutting, which as before stated is fatal in about one in seven cases. We are not told if the patient above alluded to recovered or not.

Professor BIGELOW does *not* however undertake to "remove at one sitting all calculi of any size, and whatever may be the

---

*Diseases of the Urinary Organs by Sir Henry Thompson.   Churchill, 1879.

condition of the patient." He *has* removed large calculi at one sitting, but in the essay above referred to he gives instances where he required *three* sittings, each occupying various periods—the first sitting (in one of the cases) an hour and a half, the second three hours, and the third three hours and three quarters. The assertion or suggestion to the contrary has been objected to by Professor BIGELOW himself, as may be seen from his letter published in the *Lancet*, 17th May, 1879, pages 693-4-5.

It is greatly to be regretted that in a matter so important in relation to the advancement of medical science, any false or mistaken views should, on either side, be presented to the profession or to the public.

It is gratifying however, to know that a surgeon of known experience, fairness and veracity—Mr. CADGE, of Norwich—has operated successfully on Professor BIGELOW's plan, and has reported five successful cases.* He says (among other things), " When I first became aware of the views of Professor BIGELOW on the treatment of stone in the bladder, by rapid lithotrity and large instruments, I confess that they seemed to me crude in theory and likely to lead to dangerous results in practice. A more careful examination of the subject, and a perusal of his excellent paper, have induced me to modify my first impression, and to think that in the new method, it may be found that we possess not only a novelty but a real advance in practical lithotrity."

What no doubt surprised Mr. CADGE, as it had Sir HENRY THOMPSON and others, was the impression respecting the intolerance of the bladder to instruments.

For nearly 25 years the writer has been convinced of the error of this teaching. Consequently Professor BIGELOW's announcement rather fascinated than surprised him—corroborating as it did a long cherished theory—a theory moreover successfully applied—confirming too the confident belief that

* See " Lancet," 5th April, 1879, p. 471

with apparatus suited to the purpose, the bladder or urethra need not be at all injured by a manipulation extending over one, two, or even three or four hours. The writer's experience goes to prove that the male urethra may be dilated considerably more than sufficiently to accommodate BIGELOW's lithotrites and evacuating tubes, if only the process of dilatation be very cautiously and gradually done. In prolonged dilatation, where patients are *extremely* intolerant of instrumental manipulation at first, and where urethral fever is set up, accompanied by rigors, cystitis, etc., patients have eventually tolerated dilatation to an almost incredible extent. The writer has over and over again verified this fact in his treatment of stricture of the urethra. He has cured many cases of organic urethral stricture by prolonged and unusual dilatation, that could only have been cured by internal urethrotomy. Patients have thus been spared severe and possibly fatal surgical operations.

The case of JOHN ADAMS (see Part V., Cases) will give the reader an idea of what may be done in the very worst form of case.

Professor BIGELOW's essay has indeed so satisfied the writer, that he feels he need make very little further effort in the direction of perfecting the instruments he has invented for use in these cases. He sees now more clearly than ever the advantage of one peculiar feature in his own method of treatment, viz., isolation of the stone, thus facilitating its removal, and obviating the possibility of fragments being left behind to from a nucleus for other calculi. Professor BIGELOW's process is very perfect, still, there is just a chance of a very minute fragment becoming entangled in the small sacculi of an inflamed and thickened mucous membrane, serving as a ready receptacle for additional deposition. This however cannot happen if the bladder is brought into a healthy condition before operation, which the writer's treatment effectually does.

The only obstacle he has had to encounter while using the large instruments has been the clogging of the débris between the blades, which gave him on one occasion some little difficulty.

This obstacle is entirely removed by a new lithotrite which he has had manufactured under his instruction by Messrs. MAYER & MELTZER.

To facilitate the removal of débris in BIGELOW's process, two or three lithotrites have been used in succession, the blades of one being cleaned while the other was in use. Sir HENRY THOMPSON informs us that with his light instruments he adopts the same plan.* The writer's new instrument however is so constructed that by a moveable contrivance between the male and female blades it effectually removes the débris, thereby obviating the necessity of re-introducing the instrument, if care be taken to thoroughly crush the stone before the aspirator is used.†

The diagrams which follow will, with the explanations appended, enable the reader to comprehend the working of this instrument.

DIAGRAMS.

No. 1.

DR. DAVID JONES'S IMPROVED LITHOTRITE, (shown in the margin of this page) to which is attached a débris-cleaner, the addition of which renders it unnecessary to use more than one instrument during an operation.

---

* See "Lancet," January 17th, 1880, page 79.

† MEM.:—Almost all the advantages here mentioned are referred to by MR. SMITH. See "Lancet," January 10th, 1880, page 43.

No. 2.

DIAGRAM OF THE BLADDER on the lower part of which is seen
the prostate gland, enlarged and projecting into the interior of
the bladder, with the "spray" playing upon it. Behind the prostate
is seen the rectum. The reader will understand that as the
prostate enlarges it presses on the bowel producing discomfort
there, and may be felt by the finger projecting "like an apple."

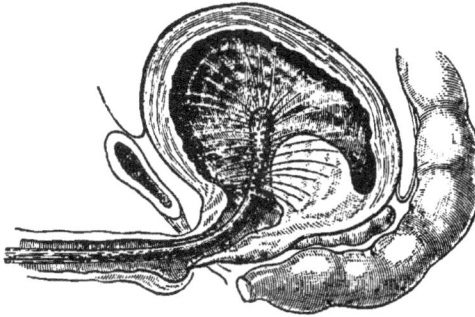

STRAIGHT AND CURVED TUBES through which the fragments of stone] are sucked from the bladder into the glass receiver.

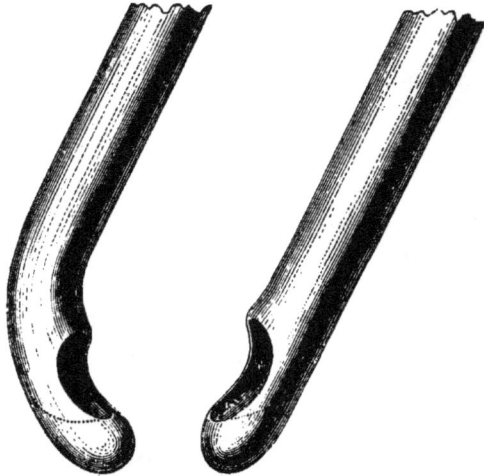

No. 4.

DIAGRAM OF THE ASPIRATOR AND TUBES, through which the fragments of stone are sucked into an improved receiver used by the writer. This plan introduced by Professor BIGELOW, is a great improvement upon the old method, where the fragments (usually irregular and flint-like) were allowed to pass through the urethra, not unfrequently tearing the mucous membrane severely, and occasionally becoming impacted in the urethral canal, from which they have had to be removed by cutting.

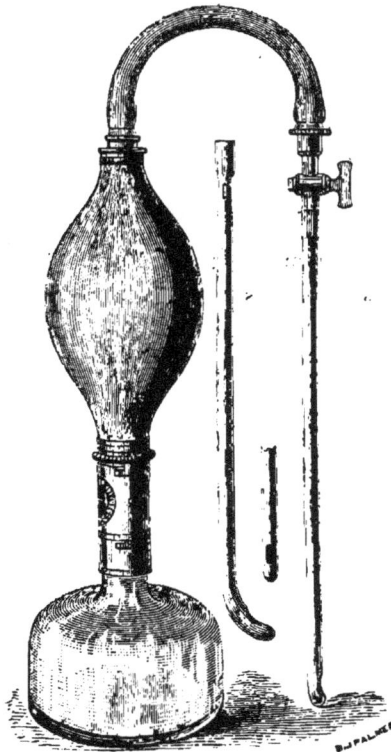

The above diagram illustrates the working of the instrument. The operator's hand is seen compressing the indiarubber exhauster, thereby forcing a stream of water through the tube into the bladder, and scattering the crushed fragments of stone in all directions. On allowing the exhauster to expand, a current of water is sucked up from the bladder, and carries with it as many of the fragments as may be drawn into the opening at the end of the tube. These pass along the tube and are finally caught in the glass receiver as shown in the diagram. This action is repeated until all the fragments are exhausted.

No. 6.

This diagram represents the search for stone with the writer's flexible sound—most advantageously usable in obscure and complicated cases.

Figure 1 represents the prostate gland enlarged within the cavity of the bladder—extending upwards and backwards, also downwards and backwards—the latter portion of the enlarged gland pressing on the rectum (figure 3) where on examination *(per anum)* it may be felt projecting "like an apple." This encroachment on the lower bowel often imparts to some patients a sensation as if the bowels "wanted moving,"—a form of expression common with them. The operator's hand is seen searching for the stone. The sound being flexible, and having surmounted the enlarged and distorted gland, is seen *above* the level of the urethral canal. During the search, it now meets another obstacle, viz., the posterior wall of the bladder. The sound, provided with a metallic tip is now rotated downwards, and (being flexible) is made to glide gently and steadily till it reaches the stone (figure 2) which is immediately recognized by the sound it yields. In some cases the stone *cannot* be made to roll from its position to the back of the bladder—hence the great advantage of the flexible sound. Figure 4 represents the ordinary sound which is stiff and inflexible. Comparing the two, the reader will readily see what difficulties may beset the surgeon when the prostatic portion of the urethra is tortuous and unyielding. It was in fact this difficulty (in J. P.'s case—see Cases) which led the writer to invent the flexible sound.

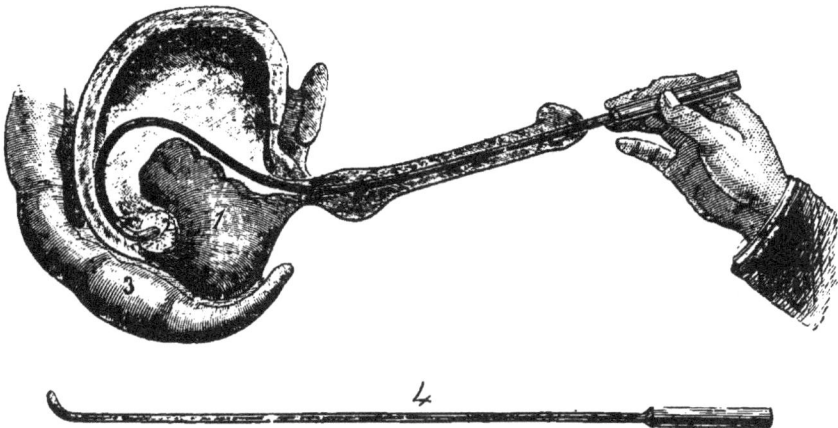

SECTION II.

DISEASED PROSTATE,
Its Treatment and Cure.

DISEASE of the prostate gland is a subject too vast to be fully dealt with in the limited space which can here be allotted to the consideration of it—and the writer proposes therefore to confine his remarks chiefly to that form of (supposed) "incurable disease of the prostate" occurring in advanced and advancing life. Many of the cases treated by the writer, and appearing in part V., show that chronic enlargement ("hypertrophy" as it is termed) of the prostate is far more amenable to treatment than authors on the subject have supposed, particularly if taken in reasonable time. Those who have written on the subject, ancient or modern, English or Foreign, and whose works the writer has consulted, all seem to have agreed in pronouncing the disease "incurable."

In a recent work written by Drs. W. H. VAN BUREN and E. L. KEYS of New York, authors of considerable renown on diseases of the genito-urinary organs, the following occurs : " In the present state of our knowledge, hypertrophy of the prostate is not curable by any means that have yet been used."*   In a still more recent

---

*A Practical Treatise on the Surgical Diseases of the Genito-Urinary Organs (1875) page 193.

work on diseases of the urinary organs, by F. J. GANT,* a similar opinion is expressed, as follows :—" In the *treatment* of chronic enlargement of the prostate, palliative measures alone have any effect—this prostatic hypertrophy, naturally incident to advancing years, being incurable." The admitted inefficacy of the ordinary modes of treatment of advanced prostatic, and some vesical diseases, suggested to Sir HENRY THOMPSON operative means for the relief of these cases. The result, however of five cases related by this able surgeon, is unsatisfactory, all the patients so operated upon having it would seem, succumbed shortly after.† These cases were however very far advanced—and the ages were, respectively, 63, 75, 50 (this patient having also cancer), 40 (a case of villous tumour) and 68. In a still more recent work,‡ Sir HENRY says :—" Medicine is powerless to diminish the hypertrophy. There may be temporary enlargement from congestion, and that you can diminish by treatment. But true hypertrophy cannot be diminished by any known means. Numerous agents have been employed, both internally and as local applications, for both of which, as might be supposed, preparations of iodine and mercury, have been particularly vaunted. And notwithstanding all that has been claimed for such agents in certain quarters, I assure you, with regret, but with the most complete confidence, that neither iodine nor mercury does anything but mischief, however employed. Besides these, other remedies as hemlock, hydrochlorate of ammonia, liquor potassæ, etc., have been tried. Nevertheless, I have simply to say that, for the present, we know no means of checking the progress of hypertrophic enlargement."

In another part of the same chapter is the following :—" The first object of treatment is to relieve the partial retention of urine

---

* J. & A. Churchill, (1876) page 214.

†Lectures on Diseases of the Urinary Organs, J. & A. Churchill, (1879) page 284.

‡ "Clinical Lectures on Diseases of the Urinary Organs," J. & A. Churchill, New Burlington Street (1879), chap. 7, 79, et. seq.

by the catheter. There are two causes which produce this retention of a certain portion of urine, and which no efforts on the part of the patient enable him to pass. The first is, as you know, the obstruction of the enlarged prostate itself at the neck of the bladder. But there is, moreover, an inability of the muscular coats of the bladder to contract and expel its contents, and it happens thus :—In order to overcome an obstacle to the outflow of urine the muscular fibres are greatly increased, and Hypertrophy of the vesical coat results to a large extent, just as the walls of the heart thicken when obstruction exists in one of its main outlets. The thickened bladder is much less distensible than the bladder of normal character, and the organ is often equally disqualified for retaining much urine or for expelling it entirely ; the cavity of the bladder being diminished, and its function as a reservoir impaired, in part by the protrusion of enlarged prostate into the interior, and in part by the rigidity of the coats as above explained."

The writer fully agrees with Sir HENRY that the "vaunted" remedies, mercury, iodine, etc., etc., do *not* cure or relieve this condition of disease. He believes moreover that they have unquestionably killed many suffering, not from prostatic disease only, but from many other diseases also. The writer himself employs neither of the above-mentioned remedies. In fact, his mode of treatment if applied in time, will dispense in many cases with the use of the catheter also. (See the "Cases"). The *cause* of the residuary urine is enlargement of the prostate gland, occasioning mechanical obstruction. To overcome this the muscles of the bladder make violent effort, and assisted by counter effort (so to speak) on the part of the abdominal muscles, the whole bladder becomes thickened in the attempt. Nevertheless, it fails to expel the urine, and in proportion to the loss of power in the expelling function of the thickened, decrepit and helpless bladder, so proportionately is the residual urine retained. In other words the enlarged prostate projects into the viscus—the already thickened walls of the bladder render the function of propulsion impossible—and hence retention of urine. This urine becomes decomposed, forming a strong, blistering, corrosive, ammoniacal fluid. And the reader will thus see why in these cases, there is such frequent desire to urinate, though only a very small quantity can be passed with each effort. The acrid, irritating urine is in contact with the most sensitive part of the bladder producing a sensation often described as "burning," "scalding," "stinging," or as some describe it, "a feeling as of boiling lead or melted iron in the bladder." This small quantity of corrosive urine is all that can be expelled, but a large quantity, varying from one to forty or more ounces is left behind as "residual urine," constantly undergoing decomposition.* The cases in Part V. shew how rapidly the lost function of the bladder is restored by the treatment adopted by the writer. Under that treatment the patient almost at once passes *more* urine by his natural effort, and the "residual urine" becomes less and less—the expulsion power of the bladder being restored. The urine loaded with mucus, pus and blood, soon becomes changed—the decomposed, offensive, bloody and putrid character of the excretion becomes substituted by urine, normal as to consistency and odour, and the patient is convinced that, despite the opinion of high authorities, enlarged and (as generally supposed) "incurable" disease of the bladder and enlarged prostate in the male, and bladder disease in the female, *can* be cured in both sexes. The case of Mrs. TOLLY, of 27, Coburn Street, Bow Road, is highly interesting. (See "Cases"). The celebrated surgeon who had unsuccessfully treated Mrs. TOLLY, on being told by her that "Dr. DAVID JONES positively assures me he can cure me," replied, in the presence of her ordinary medical attendant,— "Well, Mrs. TOLLY, all I can say is, that no *respectable* medical practitioner would *attempt* at your age to do more than we have

---

* In the case of a gardener from Elstree, the Writer's son removed 90 ounces.

done." Mrs. TOLLY however *was* cured by the writer, and continues quite well to this day, and since then has testified her gratitude to the writer by giving him a subscription towards his Home Hospital, thus helping *poor* sufferers who cannot help themselves.

The writer does not however by any means profess that he can cure *every* case of disease of the prostate that comes before him, but he does feel convinced that his method of treatment by specific local applications to the bladder and prostate is far more rational, and more likely to effect a cure in such cases than either medicine introduced into the stomach, or the ordinary kinds of surgical procedure. Moreover in every bad case undertaken by him, his treatment afforded more relief than could possibly have been afforded by any other means within his knowledge. The late THOMAS HALL, Esq. (who testified his gratitude for services rendered by the writer by a munificent gift of nearly £2,000) wrote thus :—" You told me honestly when you first saw me, my case was beyond cure, but your treatment has done more for me than any other."* That gentleman had been treated without relief by some of the greatest celebrities at home and abroad. Let the reader carefully peruse the numerous " cases " embodied in the present edition, and observe the accounts which patients themselves have given of these terrible diseases. Let medical gentlemen who cling to the opinion of high authority, investigate the whole circumstances surrounding these " cases " : most of the patients are accessible, names and addresses are at the writer's disposal, as well as those of unimpeachable witnesses who knew them in their supposed " incurable " condition ; also the names of physicians and surgeons who had attended them, and pronounced them beyond cure, before the writer commenced his treatment of them. Possibly the medical profession will say that the cases in which the writer has effected a cure were not really typical cases

---

*His attendant, Mr. Kevis, (106, Licensed Victuallers' Almshouses, Old Kent Road) still living, can testify to this.

of enlarged prostate, and that the ages of the patients do not correspond with the period when enlargement of the prostate (" chronic hypertrophy ") comes on. It cannot, however, be denied that the ages of some of the patients are precisely the age admitted by authorities as the most likely for the bladder and prostate to become troublesome. With reference to other patients who were younger, the writer merely adds that the *whole* of them, though cured by him, had been given up as hopeless by physicians and surgeons previously consulted. But the cases occurring, even in the younger persons, were in fact " typical " cases of chronic disease, and had (as the patients themselves have asserted) been previously treated by surgeons of high repute.

The writer ventures to question the soundness of what has been affirmed by many writers respecting the special pathological condition of the prostate gland ; also the accuracy of the classification of these prostatic diseases by those who assert that nothing more is to be done for patients simply because high authorities have failed hitherto, to discover, and apply more certain means of relief and cure. But is the scientific mind now dormant? Are we for the future and for ever to leave off searching for means to alleviate the most painful of human diseases because high authorities affirm that "medicine is powerless to diminish the hypertrophy—it cannot be diminished by any known means," etc., etc.? Were not similar questions propounded respecting the cure of ovarian dropsy ? The late Mr. ROBERT LISTON called the ovarian operation a " belly ripping operation." He said this to his whole class, the writer among them, and ever persuaded us not, under any circumstances to attempt it. The writer as well as many surgeons in London and elsewhere, Sir SPENCER WELLS, Bart., and Dr. KEITH in particular, can now afford to laugh at such advice.* But the writer is

---

*And see " Tumours and other Diseases of Women," compiled by the writer and published by Mitchell & Co., 13, Red Lion Court, Fleet Street.

(denounced because he does not reveal his treatment. His reasons to himself sufficient) for *not* doing so at present are given in another part of this work.

With respect to the disease itself, he believes that any real difference between one prostatic disease and another is rather in the length of time the disease has existed that in anything else. There is doubtless a pathological difference between a disease in the prostate of a man of 30 or 40 years of age, and that of a man between 55 and 65, just as there is in that of a man of 20 or 30, and of another between 30 and 40. The diseases are more or less alike, varying in degree only as already intimated, with the age, habit of body, and peculiarity of constitution. There is a great difference between a green blade of wheat three months old, and the golden mature corn ready for the sickle : still it is substantially the same, altered only by age. There is a similar difference in the pathological appearances of incipient and fully developed prostatic enlargement. The writer is convinced that the prostate gland takes considerably longer time to grow into an enlarged diseased condition than is commonly supposed by professional men. It grows, nevertheless, though *insensibly*—in other words it gives little inconvenience till the "last straw breaks the camel's back."

It is in the early stage that this as well as other diseases should be treated. For many years the sufferer may have had *intimation* or warning of the approach of the disease, but as the warning was not very urgent it has been disregarded. To guard such persons against the certain consequences of this disregard, the writer will here add some observations descriptive of the earlier symptoms usually experienced, and as usually neglected.

The first symptoms of the approach of disease of the prostate is merely an "uneasiness about the neck of the bladder." Probably during cold weather there is more frequent desire to urinate,—that is all,—and no more notice is taken of it. In the course of time the frequency becomes more urgent. Still no

particular notice is taken of it—" it is only a slight inconvenience."
In months or it may be years, the discomfort increases, and a
nightly call becomes habitual.   By-and-by the patient begins to
feel the discomfort of getting out of his warm bed.   Still not
much notice is taken of it.   He does not think it worth his
while to consult a doctor " for such a trifle."   If he did, the
probability is that his doctor would say "it is nothing," or
possibly, "it is weakness of advancing life,"—give a bottle of
medicine, and prescribe " whisky or gin," or some other more
powerful diuretic (the worst thing imaginable), and the patient
encourage himself in the belief that there was nothing the matter
with him.   The disease is all the while getting worse.   In the
course of time the patient has to get out of bed twice during the
night instead of once.   Afterwards the frequency becomes still
more urgent, and the inconvenience becomes more evident,
and eventually *pain* is substituted for inconvenience, and then
the doctor is sought.   But unless a specialist be consulted the
bladder will most probably not be examined : medicine will be
prescibed which only excites the kidneys to secrete more urine,
and which, as already mentioned in Part I., does more harm than
good—the disease slowly but surely progresses.   Patients frequently
write after the manner indicated in the following, which is
extracted from a letter recently received :—" I have had
something wrong with my bladder for a good number of years,
having to urinate more frequently than I ought, generally having
to do so three or four times during the night, and from time to
time having *great* desire to do so—very frequently every half hour
or so, and not able to do it freely, but scantily and with pain.
This latter symptom was only occasional—I mean, having to
void the urine so frequently.   I don't think, however, that for
some time past I have been able to empty the bladder completely
at once.   I have had to wait a minute or two, then try again.
Things went on in this way until about two years ago, when the
passage of the urine completely ceased, and for several hours I

was in the greatest agony, until I had my bladder relieved by an instrument which occasioned me great suffering and loss of blood. The bladder had to be relieved for two days, after which the urine began to flow again,—at first in very small quantities, but afterwards much in the same way as it did before, and it has continued much the same until a week ago, when I had another complete stoppage, attended, as before, with intense suffering, and was again relieved by the instrument, which gave great pain and more bleeding than before. After the instrument was used a few times, the water began to flow in small quantities at first ;—this has left the neck of my bladder sore and irritable, and I cannot empty it all at once. Besides the catheter, the doctor used medicine, fomentation, and poultices."

The above presents a fair account of what usually happens in these cases. Later on, or sometimes earlier the urine becomes cloudy, and still later it is found to have deposited during the night in the chamber utensil, a quantity of thick, tenacious, offensive mucus. In addition to what has already been said, there are other symptoms, often mistaken for disease, in the region of the rectum—the discomfort alluded to being attributable really to enlargement of the prostate gland, which from its enlargement presses on the contiguous rectum. Occasionally, the first intimation of enlarged or enlarging prostate occurs through sudden retention of urine, and patients are under the impression that there was nothing wrong with the organ previously. Close questioning however, quickly elicits an experience similar to that described in the letter quoted above. The truth is, that however slight the inconvenience, it should not be neglected. The symptoms should be brought under control at the outset. Great as the mistake is, it is of very common occurrence—too frequently turning out to be a *fatal* mistake. It is so, not only with diseases of the bladder and prostate, but with most other diseases—pulmonary consumption for instance, or even a case of toothache. Consumption never *begins* in the lungs.

These patients will tell you that for years before the lungs became diseased they used to suffer from constant catarrh. Susceptibility to nasal irritation has existed for probably some years—thence it travels to the throat, and the patient becomes subject to sore throat, and ultimately, after much suffering from this latter complaint, the disease gradually passes (by the continuity of the mucous membrane) into the lungs, constituting "pulmonary consumption."

It seems hard to believe that before a person suffers from an agonizing toothache the decay has been gradually progressing without pain, but eventually the decay slowly but surely reaches the sensitive part of the tooth, having occupied from seven to 12 years it may be to accomplish it. Just as the decay of the tooth may be arrested by the early attention of the dentist, so prostatic disease may, by early attention, be greatly relieved—and by the writer's treatment be not only cured, but prevented from recurring during the remainder of a long life time. Disease of the prostate takes a long time to develope, and patients rarely seek assistance until the gland has become so large as to be seldom restorable to a size where mechanical means can be dispensed with. Independently of this neglect, surgeons are themselves too much in the habit of depending on the catheter for the relief of patients, and too readily at once instruct them how to use it—telling them, moreover, "this (the catheter) is to be your doctor for life." No doubt, in the absence of better treatment, the catheter has been of great service to many, and has undoubtedly prolonged and even saved many a life. It is nevertheless a remarkable fact, that the writer has been less successful with patients who have habituated themselves to use the catheter than with any others. The constant use of the catheter without any treatment to prevent the growth of the diseased gland, or to reduce its size, allows the gland to go on enlarging,—nay, more, the writer is of opinion that the constant use of the catheter irritates the prostatic portion of the urethra and assists in increasing its size

E

until, sooner or later the mechanical obstruction becomes so large that it is impossible to have the bladder emptied without the catheter. Independently of this, the writer has noticed during his long experience that when the catheter is once commenced, even when the hypertrophy is not very great, it is with the greatest possible difficulty he can induce patients to leave off using it. The urinary bladder becomes so accustomed to its use that it jibs (so to speak)—in other words it refuses to obey the will without help. The constant friction of the catheter irritates the mucous membrane in the prostatic portion of the urethra, occasioning inflammatory exudation. This extends to the structure of the prostate, occasioning a morbid condition described by authors as a "peculiar pathological condition of chronic hypertrophy."

Corns and bunions on the toes form in a similar manner— friction from a boot occasions inflammation in the skin. This skin inflammation extends to the structures underneath, including muscle-tendons, sheaths of tendons, fat, nerves, arteries, veins, lymphatics, etc.,—all these tissues become hardened and diseased and form "roots," as they are termed, which occasionally reach to the very bone itself. These mixed tissues (so to speak) form a new morbid product, confusing the pathologist and microscopist, just as does the tissue found in chronic hypertrophy of the prostate. The truth is, that neglect and inappropriate treatment impel (so to speak) the bladder and other organs to contract bad habits. What with the catheter irritating by constant friction the urethra and prostate, and the buchu, pareira, whisky, gin, and mineral waters, working the bladder in the other direction, the poor bewildered organ, in its attempt to get quit of the fluid poured into it, becomes enlarged (hypertrophied), just as the heart becomes enlarged (hypertrophied) when there is obstructive disease. When the habit of drinking herb tea and diuretics by the quart shall be done away with, it will go a great way towards preventing hypertrophy of organs. The kidneys enlarge to an enormous size in diabetes, because they have to pump out quarts instead of ounces—in other words they become hypertrophied, just as the bladder, prostate, heart, or any other organ does if unduly worked.

PART V.

———

CASES.

———————

NOTE.—*The following Cases are compiled from the writer's note book, and wherever the language of a patient is quoted the precise words, recorded at the time, are given. This plan has been thought desirable, as being the one most usefully guiding to general readers, enabling them to see therefrom in what particular respects their own cases correspond with those here described. It may be useful also to mention that a list of the names, etc., of the several patients whose cases are described in this Part (V.) will be found in the* APPENDIX, *and that* in every *instance in which the* full *name and address are given, those patients may (by permission) be personally communicated with by any one desiring information respecting their cases.*

No. 1.

Disease of the Prostate Gland, with Inflammation of the Bladder, and other Complications.

E. S., aged 58.—March 1st, 1874.—He had been suffering since 1872 from irritation of the bladder, causing much pain and discomfort—lost flesh, looked very ill, was pale, and the expression of countenance denoted great suffering; became very irritable, was feverish and disturbed at night. A bad cold considerably increased his suffering, and brought on constant desire to urinate. He states: "I had to run in a moment when the desire to pass water came upon me; if I did not immediately satisfy the call of nature, the water came away from me by itself, attended by severe burning and spasm." The acute symptoms somewhat subsided, but left considerable irritation and desire to urinate at night, and afterwards during the day. His medical attendant, at Brentford (w.), examined him with instruments, and pronounced his case to be "inflammation of the bladder, and disease of the prostate gland." Gradually getting worse, he had to leave off his occupation on several occasions, and declare on his club; and eventually he was laid up and confined to his bed. He was invalided from March to August, 1872 (five months).

Getting little or no relief from his doctor's treatment, his late employer, Mr. Whyman, 48, King Street, Hammersmith, for whom he had worked seventeen years, advised him to go to Hammersmith Hospital. He consented, and was kindly treated there by the medical officers who took charge of his case; still, he got no relief—indeed, he became worse.

The Rev. Mr. Dribbles, of St. Paul's Church, Brentford, now interested himself on his behalf, and gave him admission to St. George's Hospital. He was examined for stone six times, by the surgeon there, who gave various opinions about his case. The treatment he was subjected to was very trying; and although an inmate of the Hospital for three months, he got no relief. In reply to his enquiries, the surgeon ultimately told him he was suffering from enlargement of the prostate gland, and, possibly, fleshy enlargement in the neck of the bladder in addition. It was, however, thought that as he got no relief in the Hospital, fresh air and good living might be of use to him; and he was accordingly removed to the Convalescent Home, belonging to the Hospital, in Wimbledon. He received every attention from the House-Physician, and was somewhat improved in general health ; but his bladder disease was no better ; consequently he returned to his occupation. He followed his work with difficulty, and was compelled to adopt various expedients to give him ease in sitting ; a cushion was placed under his right thigh, on which the whole weight of his body rested, thus relieving the painful pressure on his seat. After enduring a miserable existence, he became much worse ; and having at this time purchased a copy of the writer's little treatise on "Diseases of the bladder cured by a new discovery," he consulted him on his case. His symptoms were as follows :—(1) almost constant desire to pass water, attended with violent burning and straining, day and night, and thought himself pretty well if he went an hour without micturating ; (2) passed blood and thick discharge towards the latter end of the act ; (3) was always worse during and after exertion of any kind, and after walking, or riding in any kind of carriage—this greatly increased the tendency to micturate, but, though he had an urgent desire, he was only able to void a teaspoonful about every ten minutes ; (4) severe burning pains about an inch from the urethral aperture—this pain extended, in a modified degree, along the perineum (the crutch), until it reached the fundament, when it became more severe and constant ; a gnawing, burning pain, which gave him sensation as if the bowels wanted to relieve themselves, although such was not the case ; (5) the urinary secretion was cloudy and deposited mucous discharge, offensive to the smell and alkaline in its character ; (6) had slight stricture in the urethra ; (7) could not thoroughly empty his bladder, about an ounce of offensive urine remaining, which was drawn by the catheter ; (8) the prostate was tender to the touch, and considerably enlarged.

The treatment adopted by the writer in this case was, to use graduated catheters with a spray arrangement ; in this way the stricture soon gave way, and one obstacle to the bladder mischief was thus removed. After the third application, which was administered on the 10th of March, 1874, he expressed himself relieved. On the 28th of the same month, he states: "I passed a jelly-like substance, very slimy, not unlike a small snail; and I have been better ever since. If I could be sure of remaining as well as I am now, I shall for ever feel grateful to God for directing me to you."

April 8th—considers himself improving; urinates only twice during the night; he still passes a little blood occasionally. On introducing the catheter and spray, there is some difficulty in getting over the prostatic portion of the urethra.

April 11th—he states, "I am so delighted with myself that I sometimes cry, and at other times laugh, when I think how quickly you have cured me. I seem as if roused out of a dream, and cannot tell what I think of myself."

April 22nd—after the last application he passed a good deal of blood (not clotted), passed water only twice during the night, and three times during the day.

April 28—"I have had no return of my suffering, sir," were the first words with which he greeted the writer. "I used," he added, "to suffer from a pulling, smarting pain in the act of bending forward in my business, which was a great inconvenience to me; but this is now quite gone." Advised him to leave off treatment for a few weeks.

May 27th—he had been worse on account of not continuing the spray longer; the spray was again repeated.

Sept. 2nd—the last spray was more efficacious than former applications. He had, he says, "passed no blood for five weeks," and thought himself now "perfectly cured." Another spray, for safety, was, however, applied.

In reply to enquiries respecting his health, in January, 1875, he writes, "thanks be to Almighty God, and your skill, I continue quite well." In March, 1875, he called upon the writer, looking well, and had had no return of his symptoms; exertion gave him no discomfort. On examination it was found that he could perfectly empty his bladder. The prostate had been greatly reduced in size, but was still larger than it ought to be in health. There was no tenderness on pressure. He said, " I am as well as ever I was, and have not seen either discharge or blood for months."

Dec., 1880—he writes, " I think that you have made a *perfect cure* of me : I return you my sincere thanks for your kindness and skill. I hope you will publish my case for the benefit of other sufferers. I shall feel it a pleasure to see any one and verify your plan of treatment."

———

No. 2.

Disease of the Prostate and Bladder—Complicated with Spinal and Kidney Mischief.

T. L., aged 58.—This patient has given so graphic an account of his sufferings in the appended statement, that the writer does not consider it necessary to add anything to what is there written, —except a few words, for the benefit of enquirers.

This was a very remarkable case. The following particulars, voluntarily sent to the writer, embrace almost a verbatim copy of the letter received from the patient, with an express wish that his name and address should be published for the benefit of sufferers from similar disease.

The Vicar of the district in which the patient resides is well acquainted with the case, and also with the medical gentlemen who attended it previously to the writer being consulted upon it. These gentlemen, with others resident in the neighbourhood, were very kind to the patient during his long and severe illness, and could, if requested, corroborate his statement.

It may also be mentioned that this patient continues well up to the present time.

STATEMENT.

"In the beginning of March, 1872, I met with an accident through a boiler and coil, weighing about two hundredweight and a half, falling against me, which gave me a sudden twist that affected my whole system, and especially the lower part of back and stomach. Rest and poultices restored me in about a month. In three months after the accident I began to feel the first discomfort of a burning-like sensation in the lower part of my person, as well as a feeling as of a bunch of red-hot needles being thrust into the fundament. I could only sit on the edge of a chair; if I attempted to sit in the usual way I had a sensation as if the bowels and lower parts were being burnt with red-hot iron. When in the workshop I have often had to hold by the "vice" to prevent falling to the ground. At this time my urine flowed very slowly from me and with great pain. My bowels were also intensely confined. The water now commenced to dribble from me. I have had to remain in the w.-c. for an hour-and-a-half to two hours before I felt that the bladder and bowels had been relieved. After dragging on for twelve months under medical treatment, and quite unable to attend to business, I was sent to the seaside. I did not improve,—on the contrary I began to swell so that my things had to be let out three inches. I soon got worse again in my bladder and bowels, and the red-hot needles feeling increased greatly. I had to come home, and being in every way worse I had to shut up my shop for seven months. I was now told for the first time there was no cure for me. I passed water in the greatest agony as often as six to seven times an hour, in quantities of about half a teaspoonful at a time, which I passed with fearful straining. All this while I felt as if I could pass a gallon each time. I could not stand upright without support. I got no good from any of my doctors, so I was recommended to go to Guy's Hospital. I had to go by omnibus, which tried me sorely, as I had to get out of the omnibus four times on my way to Guy's to try to ease myself. I went to and fro in the greatest misery for four weeks without any relief. Here I was physicked dreadfully, and blistered until I was so reduced as not to be able to put one foot before the other without great pain and difficulty. It has taken me more than an hour to walk from the omnibus to my house, a distance of about 300 yards. I had to do this by holding to the railings. After five weeks' treatment I was worse than when I commenced. I now got an order from a retired physician for St. Thomas's Hospital. I attended here for four or five weeks. At the end of that time I got an order to have my medicine repeated. When I told the Doctor my water was not a bit better, and that 'repeating' was not of the slightest use to me, and after telling him all my terrible sensations, he ordered me to an adjoining room to be examined. Here he tried to pass an instrument into the bladder, but failed to do so, and had to put his finger into the rectum to help the instrument into the bladder. After examining me he said, 'There are two stones here.' I was treated at St. Thomas's for 46 weeks, and examined five times for stone by various doctors. I told one doctor I was not afraid of the knife, and I would gladly have the stones cut out. He first told me he must dissolve them, but afterwards that I was not suffering from stone. At this time I had to try to empty my bladder every five minutes, but instead of any water coming it was a violent, burning, straining, forcing, gnawing pain, extending from the bladder and private parts into the crutch, which gave me a sensation as if the bowels wanted to be relieved. My distress was so great at this time that I have had to get out of bed 30 times an hour up to one and two o'clock in the morning; after one to two o'clock I got two or three hours' sleep from sheer exhaustion. I had to place the chamber utensil in a slanting position and rested till the sudden burning came upon me, that drove me all but mad. Words cannot describe my agony; it is quite useless to try to do so. When I tell you that my hair and whiskers, which were dark, had, by this time, turned *quite white*, it may convey to you some idea of what I endured. My age at this time was 42, but I was never taken for more than 35 before my disease began. At this period people took me for a man of 60 to 65,—my continued suffering had so aged me. At this period my wife went to Mr. W., of St. Thomas's Hospital, and obtained for me admission as in-patient to the Institution. I must indeed have been very near death at this time. The doctor who had attended me so long had not the slightest recollection of me. There was now a consultation with eight of the physicians and surgeons of the Hospital. One after another tried the bladder for stone, and came to the conclusion there was none, but said I had a fleshy substance growing in the bladder. The doctors were a long time consulting about me, and it was ultimately settled I should be blistered all over my back. God only knows my suffering at that time. After twenty-four days treatment in the Hospital I was sent down to the country for sea-bathing. I seemed here to get better, but no sooner had I returned home for work, than all my old pains returned worse than ever. The sensation of wanting to pass water and unable to do so was fearfully trying. Any movement of the body, as riding or walking, increased my suffering to an intense degree. I had to be sent back to the hospital and was placed under Dr. S.'s care. He did me no good, and at last told me there was no known cure for my disease. I was now placed under another doctor, whose name I forget. This gentleman told me my case was 'quite hopeless, and that I had better go home to die.' I was sent home, and made up my mind to try Homœopathy. A medical gentleman in my own neighbourhood visited me at my own house from October, 1876, to May, 1877. He informed me I had chronic inflammation of the bladder. Besides medicine which he gave me, he recommended

a cold water compress round my abdomen and back every night
on going to bed. He was very kind and attentive to me,
but did me no good. When taking his leave of me, I was
comforted by words to this effect :—' Mr. Ludlow, it is useless for
me to come to see you again ; it would be simply picking your
pocket; your disease is perfectly incurable. You had better
prepare for another world ; death alone will put an end to your
suffering.' I then asked him, 'Is there no hope whatever?' He
replied, 'None whatever : there is no man in existence that can
cure the disease you are suffering from.' I was terribly depressed
after he left me, and felt I should like to try another
Homœopathic doctor not far from my own house. I did so, and
certainly improved a little under his treatment. I became so far
better as to be able to crawl out of bed and go down stairs
back foremost, on my hands and knees, my wife standing at my
back to prevent me from falling backwards. I had my bed
made on the couch, where I remained for four months. When
the doctor found he could do me no good he took his leave like
his predecessor, and told me my case 'was a hopeless one—
death only would put an end to my pain.' My condition now
was such that no one can describe. I had tried in all ten or
eleven different doctors, extending over a period of five and a
half years, and during the whole time I was suffering most
acutely. My disease had been variously called 'Stone in the
Bladder,' 'Gravel,' 'Painter's Colic, 'Inflammation of the
Bladder,' 'Irritable Bladder,' 'Disease of the Spine,' etc. I
looked like a poor worn-out old man between 70 and 80 years
old,—drawn double by pain and grief,—not even able to stand
unless I had support on each side of me ; indeed my countenance
bespoke that my time here below was fast coming to a close.
My toes were drawn in, and I was in all respects a helpless
wreck. Just as all earthly things appeared to be at an end, and
I had earnestly prayed that God would in His mercy help me
or put an end to my wretched condition, I saw an advertisement

of a book on 'Diseases of the Bladder and Prostate, by DAVID
JONES, M.D.,' which I procured. After reading it I became
convinced that the local treatment was more likely to cure me
than anything that had been tried. I made up my mind to
send for you, and I can confidently say that by God's blessing on
your skill I have been cured of a most painful disease by your
treatment. I have ever since been able to attend to my
business, which is that of a gas and hot-water engineer and
general ironmonger, with very slight interruption, from October
1877, and I continue well up to this date, 1880. You first
visited me in August of the same year. The first spray you
administered commenced its work speedily. The relief I
experienced was magical. I went downstairs to tea in two
days, but only remained a very short time. After two more
applications, to the surprise of my neighbours and the joy of my
wife and friends, I was able to go to Bolton House in a cart.
After my third journey to you I was able to come to you in a
tram. After the fifth, I could walk to Bolton House with slight
help. After the eighth journey I went to my workshop and
managed to do four hours' work a day. After seven weeks'
treatment I worked eight hours a day; and after eleven weeks'
treatment I was discharged cured. On two occasions since
my cure, I have had slight return of my bladder symptoms, but
an application or two soon put me right again. I have lived in
this neghbourhood for 19 years, and am well known to rich and
poor. In case of any doubt of the truth of my statement, I
can give any amount of references to satisfy the most scrupulous
enquirer. If enquirers cannot come to me I will most willingly
go to them to explain the extraordinary effects of your treatment.
I write this to you, my dear doctor, with a heart brim full of
gratitude, and trust you will give the fullest possible publicity to
this statement for the benefit of helpless sufferers. I pray,

moreover, that God will spare you long life and health to rescue from the grave afflicted sufferers as you have me.

Believe me,

Your very grateful and ever obliged Servant,

Thomas Ludlow,

Ironmonger, etc."

149, Clapham Park Road, Clapham, S.W.

No. 3.

Disease of Bladder and Prostate.

W. R., Esq., aged 68.—January 9th, 1879.—Has been a great sufferer for seven years, but considerably worse during the past two years. Treated by several of the most eminent surgeons in London, who pronounced the case to be as above designated. The writer's diagnosis shewed the prostate gland to be considerably enlarged. On being asked to urinate, he said he had "not long done so," but would try. Urged to try thoroughly to empty his bladder, he did so, and passed three and a quarter ounces. The urine was alkaline and loaded with thick mucus. On introducing a catheter, drew eight ounces and a half of very offensive urine, highly ammoniacal, the last drops loaded with very thick muco-pus, the urine was, moreover albuminous.

January 13th : first spray administered.

January 16th : considers himself improving. Slept comfortably after the first application. Urinated naturally five ounces ; drew by catheter eight ounces. Urine still alkaline and ammoniacal.

January 20th : " I am," he states, " decidedly better." Urinated naturally five ounces and a half; drew by catheter six and a quarter ounces. Urine clearer ; no muco-pus with the last drops.

January 27th : Still improving; has little or no inconvenience. Urinated naturally six ounces, which was perfectly clear; drew by catheter only two ounces.   He states : "I have not the slightest pain or inconvenience, and fancy I pass water as well as ever I did ; it is quite clear at all times.   In fact, I feel cured and sleep all night, thank God, in perfect comfort.   How curious, it seems that none of the physicians I saw did me the least good. They do not of course understand these diseases."

January 30th : Has a bad cold, is chilly, has pain between his shoulders, and feels worse in his bladder symptoms.   There is a little mucus in the urine, urinating naturally.   Urinated seven ounces naturally and without discomfort.   Drew by catheter two drachms (120 drops).

February 3rd : Recovered from cold ; urinated naturally ; perfectly clear urine ; only a few drops came by catheter.   Gave him permission to go to Bath and leave off treatment, requesting him to resume the applications immediately should his symptoms return.   The next account of him is in a letter sent on his return from Bath :—

"Brixton Road, S.W.,

24th March, 1879.

Dear Sir,—I have just returned from Bath, where I have been staying some time, and am glad to say that I feel quite well.   With reference to the disease I was suffering from for the last six years, viz., the prostate gland and bladder, I feel perfectly cured.   It is impossible for me to be thankful enough to you for your kind attention and wonderfully *perfect cure* of the disease in the short space of one month.   When I think of the many sleepless nights I have suffered from for the last two years, and the several medical men I have applied to, viz., *(here follow the names of three Physicians)* and several others, none of whom did me the slightest good, I feel more than surprised and thankful at your success.   For the sake of suffering humanity, I shall consider I can never give sufficient publicity to the cure you have effected in my case ; it fully bears out the several cases of cure reported in your 2nd edition of ' Diseases of the bladder and prostate ' in my possession. Should I at any time require the services of a medical man, I shall with the greatest confidence place myself in your hands again. Thanking you for your successful cure,

I am, dear Sir,

Yours very truly,

David Jones, Esq., M.D. W. R.

P.S.—I shall feel obliged by your informing your Assistant that I thank him very much for his kind attention."

In a letter bearing date 10th June, 1879, he writes, in answer to enquiry respecting his health, "I am glad to say I continue quite well, and have had undisturbed sleep ever since I last had the pleasure of seeing you."

A detailed account of the case is very lucidly given by the patient himself in the following letter. The circumstance which led to the letter was this : The Rev. Canon ——, who was a great sufferer from prostatic disease (since cured under the writer's treatment), had communicated with the writer requesting private references. One of the three names selected was that of Mr. W. R ——, and the following is published by consent :—

"April 16, 1879.

Dear Sir,—In reply to your letter of yesterday's date relative to Dr. Jones' cure of the disease of the prostate gland and bladder, from which I had been suffering for upwards of seven years, the facts are as follows. On retiring from my profession in Doctors' Commons some eighteen years since, I amused myself by taking walking tours in different parts of the country, and, as far as I can recollect, about seven years since, when in North Wales, a slight attack took place, causing me some inconvenience by compelling me to urinate more often than usual. This continued, and gradually got worse from month to month, affecting my rest, and compelling me to get out of bed to urinate some five or six times of a night. This increased to such an extent for the last two years that the irritation of the bladder compelled me to get out of bed upon an average sixteen or eighteen times each night. This, I found, was fast destroying my health for want of rest. I should say also I was equally disturbed in the day time, which caused great debility. During the whole of the time I was in the hands of several celebrated surgeons, who were stated to have made my complaint their principal study, the last surgeon being Mr. C——, of Guy's Hospital, who, I believe, from reports, to be the most celebrated man. None of them did me any good. I was attended by Mr. C—— for ten weeks, and, when complaining to him I was not better, he stated that I must have patience and leave it to nature and his treatment. They all gave me a great quantity of medicine, and advice what to eat, drink, and avoid. At length I was despairing and low-spirited, when one morning I was looking over the newspapers I by accident read Dr. Jones' advertised cure of prostate gland. Although I was not a believer in advertisements generally, and feeling I was getting worse seriously, I wrote to Dr. Jones. He kindly sent me one of his pamphlets, and stated he would refer me to some persons in my neighbourhood whom he had cured. This he did, and I then called upon him at 15, Welbeck Street; and in conversation I soon found that he quite understood the complaint I was suffering from. He then stated medicine could not cure, and I then put myself in his hands. The first thing he requested me to do was to urinate to empty the bladder, which I did. He then asked me if I thought I had emptied the bladder. I stated in reply I thought so. He stated he felt certain I could not do so, and he would then show me that he was right. He then introduced into the bladder a very clever little instrument, not giving me the slightest pain, and drew off no less than half a pint of water, of the most offensive kind. This, he stated, was the first cause of the irritability, and had been in the bladder getting decomposed daily by reason of

F

the bladder, through the prostate gland, getting weak and larger than its natural size. A *spray* was introduced into my bladder. I felt wonderful ease the first time, and that night I got a good night's rest, to my very great surprise. He then told me I must go to his other establishment in the Clapham Road, to his Assistant, each morning that I did not call upon him in Welbeck Street, and he would draw the surplus water which I could not discharge myself; so I arranged with Dr. Jones to see him twice a week in Welbeck Street, and the other five days to call in Clapham. This I did, and continued to do so for not quite a month; and the result, I am pleased and truly thankful to say, I feel cured. It is now going on for over two months since I left Dr. Jones, and I have had no return of the complaint, gaining strength daily, and can sleep without being disturbed, thank God. I should further inform you that from time to time, through the same instrument, after drawing my water, Dr. Jones injected into the bladder, touching the prostate gland, some peculiar gas, or medicine, and this was the means for several times of bringing away through the penis several strange looking slimy substances, in appearance like a part of a snail. I doubt not that was really the festering part of the prostate gland. Neither of his operations gave me any pain, strange to say. I find there is a great jealousy with the medical profession against him. I find also from inquiry and medical works, that not being able to empty the bladder causes a poisoning of the blood, which is very dangerous. I was not aware until lately that more than a third of the adult population suffer at some time of their life with the disease of the prostate gland, and in many cases the causes are not known. I trust I have made myself intelligible, and believe me to remain

" Yours faithfully,   " W. R.——"

" The Rev. Canon C——."

Being desirous, before going to press, to ascertain whether Mr. R—— had any return of his former symptoms, he was written to, enclosing at the same time slips of test paper requesting him to dip them in his morning urine and return them. The object in doing so was to ascertain whether there was any tendency to a return of his old complaint. The test papers—litmus—were returned well *reddened*, thus giving evidence that the urinary secretion was *normally* acid. (It may here be mentioned for the information of readers, that the urinary secretion becomes alkaline —which is abnormal—through the formation of ammonia. This is always found in most forms of prostatic urine on account of the chemical decomposition which takes place in the urine). The following letter was received from Mr. W. R——in reply :—

"Brixton, July, 1879.

"MY DEAR SIR,—I have just returned from Gloucestershire, where I have been staying for some time, which is my reason for not answering your letter sooner. I return the two slips of paper, dipped in my urine (as requested in your letter) at two different times. I am very glad to say I keep very well, never better. The Rev. Canon C—— called at my house as he promised, but I am sorry to say I was not at home. When I saw him at Bolton House he told me he felt you had nearly cured him, but that he was going to stay with you another fortnight. I consider his cure most wonderful from his statement made to me of the deplorable condition he was in when he first went to you. I cannot help mentioning wherever I go your great cure of myself, and particularly to those who know how much I suffered. I shall be very busy for some time in a matter of business, where I am a trustee, but I can manage to look in at Bolton House next Friday morning at any hour that will suit you, if you drop me a line. After that I shall be very uncertain as to my being at home, as I shall have to go to Exeter and other places.

I am, yours very truly,   W. R.——

" DAVID JONES, ESQ., M.D."

" P.S.—I feel if I had a few of your cards I could put them in the hands of my numerous friends from time to time on chatting to them."

On the 29th of July an examination of the Prostate Gland showed it to have become *considerably smaller*, and there were not more than four drachms of residual urine in his bladder, giving him no inconvenience.

———

No. 4.

Very severe Disease of the Prostate, Inflammation of the Bladder, and Senile Stricture of the Urethra.

The patient has described his own sufferings so accurately that the writer does not consider it necessary to add to the statement, except to say that the London surgeon, alluded to in the statement, did *not* examine for stricture.

STATEMENT OF GENRL. SIR F. H.

August 20th, 1880.

" In grateful acknowledgment of the wonderful and successful treatment by Dr. JONES, 15, Welbeck Street, Cavendish Square, and in justification of the efficacy of his new discovery, I am very desirous of making the following statement of my case as public as possible :—

" I was taken ill on the 5th February, 1880, with disease of the bladder, and enlargement of the prostate gland, was attended by my country surgeon who ordered me warm baths, and prescribed the usual medicines in such cases, but to no effect. I became worse and worse, the constant desire and difficulty of emptying the bladder, attended with violent burning and straining, continued day and night. This pain extended along the perineum until it reached the rectum, giving me constant desire to relieve the bowels each time I relieved the bladder. The urinary secretion too was cloudy, depositing a thick yellow ropy discharge, offensive in its odour and alkaline in its character. I had also stricture in the urethra. At last my strength and appetite failed me, and

being considered in a critical, dangerous, state, the best surgeon
from Northampton was called in (want of power to relieve the
bladder having set in) who introduced with great difficulty (after
trying various sizes) a small catheter, which was left in for six
consecutive hours, causing an abrasion (found out afterwards by
Dr. JONES) the agony of which I can describe to no one, nor the
blessed relief when it was removed, and a warm bath administered.
This relief however did not last, and finding the Northampton
surgeon's treatment was of no avail, a celebrated hospital surgeon
from London (noted for having published and lectured on
" Diseases of the bladder ") was called in. He stated that I
could not possibly have a stricture, as it ' never set in over the age
of 60.' He ordered nourishing diet and milk, also suppositories to
be administered at night, the warm bath and medicine, and
steaming the parts affected. But all these gave only temporary
relief, my strength and flesh left me, I had constant suffering, little
sleep and next to no food being taken the very sight and smell of
the nicest dish causing nausea. I must also mention I suffered
from continual sickness of the stomach, accompanied with
occasional vomiting. The surgeon now told me I might apply
two or three suppositories a day, they would do no harm, and
were better than taking opium through the stomach, as they act
directly on the parts affected by disease, without injuring the
stomach, and in this state (my country surgeon having met with
an accident which laid him up) I was left.

"I had now been 4 months and 10 days ill. My nurse in her
excessive grief and despair at my dangerous state was moved by
Divine Providence to search in the "Standard" if she could find
any medical treatise, or the name of any physician that might be
of service to me. She succeeded, and, without saying a word to
me, got Dr. DAVID JONES' treatise on " Diseases of the bladder
and prostate,"—read it first herself to be quite sure it would be of
service to me. She then read it out to me, and entreated me as
I valued my life, to place myself under Dr. JONES' treatment. I

consented, and truly I was in a dying state when first I presented myself to Dr. JONES, and dried up like a piece of parchment by opiates. Within six weeks (thanks to Divine blessing and his able treatment attended with very little pain and suffering) I am perfectly restored to health and vigour. The bladder now empties itself completely, whereas when I first went to Dr. JONES he had to draw off eight ounces and a half of decomposed urine. I can solemnly declare that I have not for years enjoyed such excellent health as I am doing at present. I beg to add that my age is 74.

"Should anyone desire to have further particulars of my case, I shall be most willing to communicate with them privately, for which purpose I subjoin my address.

"GENRL. SIR F. H."

[Here follows the Address which the Writer will forward to any enquirer.]

"To Dr. DAVID JONES,

"15, Welbeck Street, Cavendish Square, W."

February, 1890.—Up to this date SIR F. H. has experienced no return of his former disease.

No. 5.

Chronic Enlargement of the Prostrate Gland and other Complications.

The Rev. Canon C——, aged 66.

The above gentleman consulted the writer on the 14th May, 1879. The case is reported just as it was entered in the Case-Book.

1.—Canon C—— states that he has been suffering from urinary inconvenience "off and on for twenty years," but the trouble did not necessitate medical treatment, till ten years ago, when his suffering became more severe. His discomfort increased gradually and from that time he became much worse.

2.—The climax of his present suffering he dates to have occurred on the 1st day of September, 1878, when in Bristol—he had total suppression of urine. An eminent surgeon, Fellow of the Royal College of Surgeons, Mr. C., was sent for, who treated him very successfully for his immediate discomfort by drawing his water from time to time. In five days he gradually regained power to urinate naturally.

3.—He still continued to suffer from inconvenience in his bladder, and was accordingly sounded for stone, twice by his surgeon and once in consultation with a physician at Newcastle-on-Tyne—one who may be said to stand at the head of the medical profession in Sunderland. It was said by this physician that Canon C—— had enlargement of the prostate gland.

4.—Besides his urinary discomfort he suffers from orchitis for a second time—a disease which occasionally troubles patients who suffer from disease, or enlargement, of the prostate gland.

His present symptoms are—(1) Constant desire to urinate night and day, but is worse during the night. (2) Constant pain in the urethral canal for about an inch down,—this pain is worse after urinating. (3) The urine is cloudy and mixed with blood and mucus—it adheres to the chamber utensil, and is "thick and ropy, and difficult to detach,"—it smells offensively and has an ammoniacal odour and an alkaline reaction. (4) He passed in my presence half an ounce of urine, which was all he could pass. (The writer drew by catheter eight ounces and a half of fœtid urine loaded with mucus and blood.) (5) Complains of great discomfort in the rectum, which is as troublesome to him as the bladder is. Treatment was administered.

On the 17th May a second consultation and application of treatment, when he stated "I am better." On being asked to specify in what way he was relieved he replied—"I have lost the severe burning pain I had when urinating to a great extent, and also the rectum discomfort which was very distressing to me—the blood is also gone." He was asked to urinate—he did so to the amount (by measure) of two ounces—the writer drew by catheter five ounces and a half. Two days previously he could only pass half an ounce, and the catheter brought away eight ounces and a half of very offensive urine loaded with blood and mucus.

May 19th.—The urinary secretion is clearing—the pain is considerably abated—urinated naturally two ounces of water—drew by catheter only four drachms (half an ounce) of urine—a great contrast with the amount drawn on the 14th of the month—drew by catheter five drachms of urine.

June 7th.—The patient states : " I am improving. I slept four hours and a half last night without inconvenience—the pain is quite gone. I go in the day-time occasionally from two to three hours without inconvenience—in fact there is a decided improvement—the pain in the rectum, which used to be very trying, is now quite gone." Urinated naturally two ounces having not long since urinated, drew by catheter two drachms.

June 15th.—To day he entered the consulting room with a firm step, and without a stick or umbrella, which had hitherto supported him. He said "I feel so well that I really do not think it necessary to enter your establishment. As far as my feelings go, I feel I am *quite cured*. I slept without interruption several hours and awoke refreshed, which had not been the case since my illness." He then said—" I hope, Dr. JONES, for the sake of suffering humanity, you will not let your secret die with you. How curious it is that medical men are so incredulous. My disease is said to be incurable. Your enemies are among the profession."

July 13th.—Continues very much better, but off and on has inconvenience during the night. Examined his urinary secretion, which was normally acid. There is not the slightest trace of cloudiness or albumen in the urine. He empties the bladder naturally and without discomfort—is occasionally inconvenienced with irritability after exertion. Examined him for stone but found none.

July 15th.—Looks well in the face. Walks straight, has lost
the tendency to stoop and the old man-look which he had when
he first came for treatment. He states—"I feel so much better
in my general health. My legs used to give way under me, but
now I feel firm on my legs. Before I came to you I had given
myself up altogether, but now think if I could get rid of the
remaining slight inconvenience I might be useful for the remainder
of my days. I used to have total inability to urinate, but do so
now without any discomfort. The forcing pain I could only compare to labour pain in women. This condition had existed off
and on since September, 1878, and without the catheter I could
not pass a drop of water. The constant pain in the perineum and
rectum has left me for some time. The blood and pus-like matter
used to scald me terribly while I urinated. This is now quite
gone, and I feel in all respects cured excepting that I pass water
a little more frequently than I used to do, which I suppose is not
uncommon to persons of my age." The urine is clear, of an acid
reaction—has no trace of albumen and has a specific gravity of
1·020. Went back to his family.

February 6th, 1880.—Comes to the writer after an absence of
ten weeks. There is no return of his painful symptoms, and with
the exception of slight frequency of urination thinks himself well.

February 28th, 1880.—Comes to the writer on a visit—stays a
night. Examined the prostate, which is considerably reduced in
size. The medium portion appears quite normal. He states—
"I feel quite well, excepting a little feebleness of age, and intend
resuming my duties the first Sunday in July, after an absence since
the 30th September, 1878 (one year and nine months).

September, 1880.—Comes to report himself on his way to Kent
to see a branch of his family. Has been doing duty for two
months, which he has continued "without any distress at all."
He adds moreover "I journeyed yesterday 300 miles without
any discomfort, and have to-day ridden in omnibuses without
any inconvenience."

The following letter represents his condition at the date it was written :—

July 8th, 1880.

"DEAR DR. JONES,—I send a few lines which will not impose upon you the trouble of a reply, and which will I think be satisfactory to you. There was a strong feeling here amongst the medical men that no good would result from my London treatment, and that the actual *reduction* of an enlarged prostate is simply an *impossibility.* My own surgeon was in church the first time I preached, to make his observations on my appearance and power. A few days after he called, and expressed his surprise at my healthy looks and power of voice. Some conversation followed on the treatment I had undergone and the gradual abatement and disappearance of my painful symptoms. I told him that if it would be any satisfaction to him he was quite at liberty to examine me and judge for himself. He called for this purpose this morning, and after a careful examination (per *anum*) said that the reduction was wonderful, that the prostate used to project into the rectum "like an apple," and that now the enlargement is not more than many, if not most, men of my age experience. This testimony of a competent and prejudiced judge, giving an independent opinion, and able to compare my present condition with the past from personal investigation, is very satisfactory to me and will be pleasing to you. And it must be remembered that Mr. M. (the surgeon) had never known me at the worst, for in the four and a half months which intervened between leaving Bishop Wearmouth and coming to you, my ailment had no doubt progressed considerably. My own daily experience, and to-day's confirmation of my state have given me a fresh consciousness of how much under God's good Providence I owe to your skilful treatment.

"With kindest regards to Mrs. Jones,

"Very sincerely yours, WM. C."

No. 6.

Disease of the Prostate and Bright's Disease of the Kidneys.

W. F. L., aged 65, widower.

October 1st, 1877. —The history of his sufferings dates from the year 1847. In that year he was for eleven weeks an inmate of Guy's Hospital, during which time the catheter had to be used repeatedly, under the influence of ether. On another occasion some considerable time afterwards, he sought relief in King's College Hospital; but after three months treatment with catheter and other means, he left without deriving much relief. He remained some time without any treatment, got worse, and went to St. Peter's Hospital, where he was treated by the catheter and other means, with little or no benefit. He now tried a local medical practitioner, who after a long course of treatment without benefit recommended him to enter Guy's Hospital again. He complied and on this occasion was discharged in three weeks much better, but still not free from urinary trouble. Some time after this his disease altered its character and assumed a much more severe form. The patient himself described his suffering as being of " greater poignancy." He had been constantly under medical and surgical treatment for twenty months previously to his consulting the writer, and during which time he had become much worse. The writer's note book records the following :—" He has constant desire to pass water night and day but is much worse during the night. He says, " I used to be in and out of bed continually, until at last I found it useless to try to lie down, and therefore sat in front of the fire wrapped up in blankets the whole night. I have taken so much sleeping medicine to relieve me that I feel perfectly stupid all night. When I am seized with a spasm, the forcing is so severe that I am afraid something will give way in my bladder. The more I force the less the water comes ; and I have to wait in agony till the spasm is over. When I pass water I can only ge away about a tablespoonful at a time, which is attended with great agony." He further said " The medicine taken made me so bewildered that I hardly knew whether I was recovering from the effect of the last effort to pass my water or from a desire to pass it afresh. I strain so much that I am afraid of forcing my bowel down : there is a sensation as if there was a large motion impacted, but it is not so."

In treating his case, the catheter drew urine to the amount of $17\frac{1}{2}$ ounces. The urine was loaded with thick muco-purulent-like discharge—a discharge resembling pus and mucus mixed : it was alkaline and albuminous. On examining the prostate by the rectum, it was found to be considerably enlarged and hard, the right lobe larger than the left.

After treatment, and on the 10th October (1877) he stated, " I have been able to go to bed and sleep comfortably ever since I last saw you." On this occasion he urinated naturally three ounces. Drew by catheter $14\frac{1}{2}$ ounces which was cloudy and thick towards the end. On the 18th he said : " I have been able to go to bed ever since the second application on the 4th October." October 22nd, continuing to improve, he said " Before I saw you I could not pass more than about a tablespoonful of urine at a time, and that I was constantly called upon to do. On the 17th of this month I was able to pass three ounces of urine the first thing in the morning, four ounces about one o'clock in the day, and seven ounces at five o'clock in the afternoon. Each act was performed naturally, and without straining or other inconvenience." He had however now taken a severe cold which somewhat renewed his suffering. On this occasion (22 October) he urinated naturally eight ounces much clearer than the last ; no traces of albumen ; the urine slightly acid. Drew by catheter four ounces. November 1st his cold is well. He says : " I passed water yesterday at intervals of about five hours, four times, and once in four hours during the twenty-four hours without any discomfort. Indeed there was a grateful feeling of satisfaction after each act, and I passed in quantity as nearly as I could judge from six to seven ounces each time. It is now three weeks and three days since I first saw you. It is truly wonderful the number of good doctors I have consulted; yet they knew nothing of my disease, or I am sure they would have cured me." He urinated naturally seven ounces and threequarters; drew by catheter only two ounces.

On the 31st January, 1878, he called to report himself and his language was noted as follows :—" I am quite well, sir, I keep my water the whole night and sleep like a babe. God bless you! I shall now begin work again as ship-carpenter. I have done nothing since the 14th May, 1876." On this occasion he urinated 11¼ ounces of perfectly clear urine. Drew by catheter sixty drops of urine. He appeared perfectly well. The large prostate had gradually reduced in size during the course of treatment but was still large—slightly larger than normal for a man of his age. The portion of gland which had encroached on the bladder and occasioned the serious inconvenience was undoubtedly *perfectly cured*. At all events, the patient has not had the *slightest* inconvenience since he was discharged as cured in January, 1878.

On the 7th February, 1878, he called again and said : " I have brought you a testimonial which you can do as you please with. I have got a berth as ship-carpenter in the "Shannon" bound for Melbourne, Australia. God be praised for directing my attention to your advertisement and sending me to you. What misery and pain as well as money it would have spared me if I had known of you sooner. Good-bye, dear doctor, and may God bless your efforts and prolong your life to open your hospital for the good of others." The original testimonial alluded to was mislaid. The patient being written to however for another, the subjoined letter was received in reply :—

"Upper Market Street, Woolwich,
"July 23rd, 1879.

"DEAR SIR,—I have not kept a copy of the testimonial but I send you from memory a summary of what I think I wrote. I first applied to Guy's Hospital in 1847 and was there eleven weeks during which time the catheter was used under the influence of ether. I then went to King's College Hospital some considerable time after I was in Guy's, and was there for three months under the usual treatment by catheter, but did not derive much benefit. I then went to St. Peter's Hospital for six weeks where the catheter was also used. I soon got tired and then employed a local doctor, who honestly confessed he could do nothing for me and advised me to go to Guy's to another surgeon. On this occasion I was an inmate for only three weeks, and derived some little benefit. After a short time the pain returned with greater poignancy preventing me from going to bed of a night for many months. I applied to you on the first of October 1877, when you withdrew from the bladder $17\frac{1}{2}$ ounces of thick glutinous urine of a most offensive character. From the first application you gave me I steadily got better. I have continued well ever since; and now instead of having to *attempt* to pass water every ten minutes in pain indescribable, I only require to do so three or four times a day in perfect ease scarcely ever being disturbed of a night I make my case known to all I find similarly suffering, and hope you may be spared many years to assist suffering humanity.

"I am, dear Sir, with grateful thanks,

"W. F. L."

"P.S.—The names of the medical gentlemen that attended me are as follows :—[*Here follow the names.*]

Up to the end of September 1881 the above patient continued quite well.

No 7.

*Chronic Disease of the Bladder, with Stone in the Bladder.**

E. B. consulted the writer in November 1874.—His appearance gave evidence of great physical suffering—complexion sallow—face careworn—tongue red and irritable—pulse small, weak, and 103 per minute—was feeble generally—very irritable and unable to take much exercise without intense suffering and exhaustion. After he had urinated (which he did with considerable pain and straining), a catheter was introduced and drew about an ounce of fluid which the bladder was unable to expel. The prostate was enlarged, but more on the left side, and tender to the touch. The urinary secretion was highly alkaline, albuminous, cloudy, ammonical and deposited thick ropy mucus. Specific gravity 1·006. His case was considered a very grave one, and after having been under eminent *Allopathic* and *Homœopathic* physicians and surgeons without relief, he was advised to come to the writer's establishment and receive his personal attention. He consented to do so, and the new treatment was soon commenced. After a few applications he expressed himself somewhat relieved, but unfortunately he took a severe cold which went on to a low form of bronchitis, and continued upon him nearly the whole time he was under treatment. The writer has every reason to believe that this circumstance delayed his recovery. However he slowly improved, the pain lessened, the urinary secretion assumed a clear aspect, the mucus disappeared, and with it ammoniacal odour and albumen, the urine soon gave the normal acid reaction, and the specific gravity became 1·023 instead of 1·006. The patient was under treatment thirteen weeks. Six months after he left London, viz. July, 1875, many of his old symptoms returned which greatly alarmed him and he soon presented himself again for advice. He then informed the writer : "I continued well until I

---

* The patient affirms he is cured, but the stone still remains.

drank a bottle of wine with a friend one evening : the next day I rode in an open carriage—it was raining the greater part of the time. I took a bad cold and my symptoms returned." On this occasion he remained under treatment about ten days. At the expiration of this time he came saying " I am perfectly well again ; in fact I thought of returning home without seeing you. I put myself to a severe test yesterday by being on my legs nearly all day without experiencing the slightest inconvenience." He returned well satisfied with himself and the writer was gratified at the success of the new treatment. The following letter received from him fully explains his case :—

"November 13th, 1875.

" MY DEAR SIR, I have pleasure in forwarding you the following particulars of my case of which I beg you will make any use you please. From a *very early* age I suffered inconvenience and pain in passing urine, and at times I remember having difficulty in emptying the bladder at all. I went on thus until the age of 14. When at school I lost all control over the urinary organs being unable at times to pass urine when strongly desiring to do so and equally unable at other times to retain it one moment after feeling the desire to pass it. None but those who have suffered the like can have any idea of the pain I endured at that time, and it would be futile to attempt to describe it, but (like many another schoolboy who rather than run the risk of being called a 'sham' will *grin* and *bear* anything) I made no complaint and consequently received no medical treatment of any kind. The attack nevertheless gradually disappeared, and I was not *seriously* troubled again for some five years when owing as I supposed to the effects of a severe cold, I passed from the bladder large quantities of blood both liquid and coagulated together with mucus. I had advice from a physician of considerable reputation but to little purpose. After being in this condition for about nine months this disease suddenly disappeared but returned again after a month or so with more fury than ever. It would be impossible

to describe in detail the condition of the urine at this time. It
was foul in the extreme containing blood mucus and sediment of
various kinds to an almost incredible extent and emitting a most
disgusting odour. The pains became very great; intense throbbing at the neck of the bladder; urging cutting and burning pains
in the urethra, stinging at the extremity of the penis, gnawing in the
groin, soreness inside the thighs, aching over the lower part of the
abdomen, continual desire to urinate and intense pain (which
I cannot describe on paper) while doing so. I again sought
advice, and after about six months the blood disappeared but I
continued to pass great quantities of mucus and the pains
increased rather than otherwise. Thus I went on for about four
years when I sought the advice of another physician the late
Dr. R——— (Homœopathist) but did not get relieved to any
appreciable degree. Another year passed: I was no better and
totally unable to do anything that required the least exertion. I
then obtained a copy of your pamphlet on ' Diseases of the
Bladder,' and eventually placed myself under your care. I
persevered with your treatment until the condition of the urine
became *healthy*. I then discontinued for a time but finding that
some of the pains still clung to me I again sought your advice,
and am thankful to say that after about ten days further treatment
I became thoroughly well. I then began to enjoy the pleasurable
sensation of existence without pain—a sensation which I assure
you none can appreciate but those who like myself have suffered
inconvenience and pain almost constantly for many years. Should
any "Didymus" require confirmation of the above facts do not hesitate to give him my name and address and believe me to be,

" Faithfully yours,

" ——— ———."

Some time after E. B. was cured he called on the writer
complaining of a sensation as if something was rolling about in his
bladder, but he had no pain or urinary discomfort. Suspecting
the presence of calculus he was with difficulty examined by the

smallest silver catheter] procurable when a stone of considerable size was detected. One of the writer's assistants who was present at the time verified its presence. The patient was *strongly urged* to have the stone removed but refused to assent on the plea that as he had "no pain or serious discomfort" he did not care to have it interfered with. Some considerable time after this he was written to and the following reply was received :—

July 15th, 1879.

"My Dear Sir,—Thanks for your note. I cannot make up my mind that I have stone—it seems to me that if I had I should necessarily be more plagued than I am. However it is a matter that can be easily corroborated some day when I have the opportunity of paying you a visit. Just now for various reasons it would be difficult for me to get away. With kind regards, hoping to see you in health when the fates permit my coming to Welbeck Street.

"Believe me, &c.,

"E. B."

NOTE.

One element in the treatment of the above case though more particularly referred to in a preceding part of this present edition, nevertheless seems to deserve special mention here inasmuch as its application so greatly contributes to the success attained in such cases. The treatment alluded to has reference to the *preparatory* treatment adopted by the writer—a treatment which renders the bladder remarkably tolerant of the existence within it of a stone, and which as the writer has found diminishes to a minimum the chances of death under subsequent operation in even the severest of cases, and gives to the writer the utmost confidence in the application of his own peculiar plan—peculiar that is in the sense of its being wholly distinguished from the methods of treatment generally applied.

The case too (of C. I. W.) brought under notice in the following observations and which was similarly treated furnishes a further illustration (out of many that might be adduced) corroborative of the foregoing remarks.

This patient came to London to be treated for disease of the prostate gland. While under the writer's treatment a large stone was detected in the bladder and the patient was accordingly urged upon to have it removed. He repeatedly told the writer in the presence of other medical gentlemen, " I am so much better from your treatment that I care not for the stone. You have cured my prostate, and I don't care for the stone as it does not inconvenience me." The following letter shows this patient's improved condition when he left London :—

" August, 1880.

" I, C. W. I., Postmaster of ———, Northamptonshire desire to testify to the great benefit I have received from the treatment of Dr. DAVID JONES, of Welbeck Street, and 192, Clapham Road, London, and I wish for the benefit of others who are suffering from the most distressing complaint that mankind can be afflicted with and from a desire to manifest my gratitude to Dr. JONES to make this statement, viz.—That after having been afflicted from February, 1880, to the middle of July of the same year from enlargement of the prostate gland, and all that time being under my own family doctor who could give me no relief only what I could obtain from morphia which was fast taking away my life, and being told that I must patiently bear it and get as much fresh air as I could, as there was no cure for it; my life became a burden to me. I happened at this time to see Dr. JONES' advertisement, and wrote for one of his books on ' Diseases of the Bladder and Prostate Gland,' and although against the advice of some of my friends I felt determined to try what the doctor could do for me. Accordingly I applied to him and was admitted into his establishment at Bolton House. From the very first application of Dr. JONES' treatment I received some relief. After the second considerable improvement was manifest, and after being with him a month and under his treatment twice a week, I have nearly lost all the bad effects of the enlargement of the gland and I hope in a day or two to return home perfectly cured with a heart full of gratitude to the doctor for having (through the blessing of God) restored me to my family and friends again. I shall be happy to answer any inquiries respecting Dr. Jones, who has restored me to comfort and happiness.

Since the above date Mr. C. W. I. has had the stone removed with great relief. Fears are entertained however that there is another stone *impacted* in the bladder.

No. 8.

Chronic disease of the bladder and prostate.

Mr. R. C., aged 29, single, residing at Stanwell, near Staines, came to consult the writer. Being himself absent, the patient on the 29th June, 1876, consulted the assistant, who attended him till the writer saw him, viz., till the 12th July, 1876.

He stated that two years previously to his coming he was suffering greatly from pain in the back and bottom of his spine, constant desire to pass water and great pain at the tip of urethra extending down the canal for some distance; also intense pain during defecation, *i.e.*, when the bowels were moved. His club doctor had attended him for " prostatitis " without affording any relief. He then went to St. Peter's Hospital—was there three months as an out-patient, and for a similar period as an in-patient. Two of the surgeons there said that he had enlargement of the prostate; another that he was suffering from paralysis and irritation of the bladder. Deriving no real benefit either from medicine or from catheters (kept in the canal for three hours at a time) he was told that no more could be done for him, and was advised to drink gin and water. He left the hospital continuing to experience great suffering. Shortly afterwards and by recommendation he went to St. George's Hospital. Whilst there as an in-patient five surgeons examined him, and he was treated for abscess in the neck of the bladder and prostate. No satisfactory result following he was advised to go to the Wimbledon Convalescent Home but declined to do so.

The spray treatment was applied for the first time and special instructions given regarding diet and for improving the general health (an important item in all cases).

On the 19th July, 1876, seven days after the writer first saw him, and after twelve days' treatment by his assistant, he wrote : " I am better ; the pain in the neck of the bladder is not so bad ; the intense pain I had when my bowels were opened is nearly gone." On the 29th July, in further consultation he said (his words being recorded at the time) : " I am very much better ; the coldness and pain is substituted by a feeling of warmth and comfort. I get a natural inclination to pass water and pass it without pain. Before I came to you I had a constant desire to pass water, and when I tried to relieve myself I could only pass a few drops at a time, which made me feel as if liquid fire was passing. I can now sleep seven or eight hours and wake in comfort ; I pass water about five times a day and plenty comes, instead of a few drops every quarter to half an hour night and day. Formerly during the first hour at night, I had to get out of bed about ten times ; during the second hour seven or eight times ; the third hour about six times, but towards the morning I generally got better. When I retired to sleep last night I felt as if I had never had anything the matter with me, and I heartily thanked God I had come to you."

On August 14th he states that for some reason he " is worse ; great pains in the groin, more inconvenience in the fundament, an itching on the right side of the urethra and a burning on the left side." His skin was found to be hot, his circulation too rapid—he had taken cold. Appropriate remedies were applied, and he gradually improved in health and strength, and ultimately be quite recovered. In a fortnight afterwards he returned to his occupation which he had not been able to follow for a long time. Satisfactory intelligence was received—his disease never returned. On the 23rd February, 1877, he wrote as follows :—

> " Stanwell, near Staines,
> " February 22nd, 1877.

" DEAR SIR,—I write to tell you that I feel quite well, and shall not require any more treatment. I feel better than I have done for the past two-and-a-half years : I consider you have done me more good than all the doctors I have been under. You can make any use of my name you think proper, and I shall be most happy to answer any questions or letters sent to me respecting the wonderful cure you have made of me, after being under many medical men and being in different hospitals. I return my sincere thanks and shall for ever feel grateful to you. I will come up in the course of a week or two and then I will call and report myself.

> " I am, dear sir,
> " Your ever grateful patient,
> " R. C.

" To Dr. Jones."

As two years had elapsed since the above letter was received, he was written to in the early part of April (1880), and the following reply was received from his sister :—

> " Stanwell,
> " 17th April, 1881.

" DEAR SIR,—My brother having a bad finger, wishes me to write for him. I am happy to say he is quite well; thank God he has not had the slightest return or symptoms of his complaint. He wishes me to say that he shall be most happy for you to publish his case and use his name, as he is indebted to you, and

you *only*, for his extraordinary cure, and the very good health
he now enjoys, and that he can never be sufficiently grateful to you
for your successful treatment of him.

"I remain, dear sir,

"Yours truly,

"S. C.

"To Dr. Jones."

The subjoined is from the copy of a letter handed to the writer
by the Rev. Canon C——-, who, previously to consulting him for
himself, had made enquiries, through a friend, as to the genuineness
of the above case, as also of the case of S. H. T. (see Cases).

"S——— Vicarage,

"17th April, 1879.

"DEAR SIR,—I very gladly went this evening to see my
parishioner, R. C., respecting the subject of your letter.   He told
me that it was under Dr. JONES' treatment that he was cured of
his painful disease, and that all the particulars contained in his
statement were quite true.   He is now perfectly restored to health
and strength and daily follows his arduous trade of a blacksmith.

.   . Hoping that you may soon obtain benefit and yet awhile
be able to return to your people in good health.

"I am dear Sir,

"Faithfully yours,

"R. P. B.

"The Rev. Canon C——."

————

No. 9.

*Stone in the bladder relieved—apparently cured—while the
calculus was still in the bladder.*

I. B., aged 61, married.

March 9th, 1878.—The history of this case led the writer to
suppose that it might be stone or prostatic disease, or both.   The
following is a summary of the writer's notes of the case :—

(1) Has passed reddish sand with his water for a long time. (2) Has suffered from pain in passing water for ten months, gradually increasing in severity. (3) Passes about a teaspoonful of water every ten minutes during locomotion ; but does not think the pain worse after urinating than before the act. (4) Has occasionally seen specks of blood imbedded in the thick ropy mucus which is always deposited in the chamber utensil. (5) Has had considerable forcing discomfort in the rectum from an early period of his suffering. (6) Urinated in my presence about a table-spoonful of thick urine, which appeared to fall perpendicularly from the meatus, as happens in prostatic disease. (7) Had been told by the physicians and surgeons previously consulted that he had stone, and operation advised. This frightened him. He complained of the pain he had experienced from various examinations he had undergone. As there appeared some doubt about the case, and the patient being very nervous about being " sounded," a mild spray was administered, and he was requested to come three times a week.

March 11th.—On entering the consulting room he said, " I am better, sir ; last night was the best night I have had for four months. I have been out of bed only twice during the whole night since I last saw you." Introduced the catheter, and drew about two ounces of urine and gave a second spray.

March 16th.—Drew by catheter half an ounce of urine, which was much clearer. He states : "I have seen no blood since the first application."

April 1st.—Has rapidly improved since last consultation. He states : " I have only had occasion to get out of bed once during the night since I was here. My water is quite clear. There is no mucus in the chamber ; it used to be thick and just like a pancake at the bottom."

April 6th.—He entered the room with a nimble step and a cheerful countenance, saying, " This is the best week I have had for the last twelve months. I can hardly believe it. I noticed the improvement particularly on getting upstairs from the kitchen. I used only to get up one leg at a time. I had to drag one leg after the other, so to speak. The water is quite clear. It is such a comfort to be in bed all night without being disturbed."

April 15th.—Drew about half an ounce of urine before applying treatment. He states, " I continue to improve. I went to bed at ten, and got seven hours' sleep right off. When I passed water I did so without any pain; and I feel I shall soon be well now."

June 19th.—The patient came as usual, apparently all but well. On this occasion a firmer instrument was used with which to administer the spray, the patient being in a standing posture. For the first time a heavy stone was apparently felt grating distinctly against the catheter. This led to a more careful examination by a metallic "sound"; the result showing unmistakeable physical evidence of stone. "Click" after " click" was elicited audible to the bystanders and to the patient himself. At this period the writer was very desirous of curing stone in the bladder by dissolution and took considerable trouble to ascertain the exact measurements of this one in all directions. This was no doubt trying to the patient—the irritation in the bladder annoying him. Before this he had appeared perfectly well satisfied with the treatment. Now he became vexed at the discovery of the stone and never came again. Some time afterwards his son informed the writer that the stone had been removed by another surgeon.

No. 10.

Spermatorrhœa and urethritis, followed by gonorrhœal prostatitis and cystitis.

W. P. aged 21, single.

October 28th, 1875.—(1) Had been suffering for four years (but worse during the past two-and-a-half years) from most distressing symptoms which resisted every kind of treatment—had consulted some of the most eminent physicians and surgeons in London in vain—had been guilty of a common indiscretion for eight years Four years ago he abandoned the habit for a time. (2) Afterwards however contracted gonorrhœa of a severe kind which for six months resisted ordinary remedies. (3) Soon after, considering himself well, he began to suffer from bladder discomfort. (4) After this he became troubled with nocturnal discharges which occurred frequently twice and more during one night. (5) Constant bladder irritation now came on, which worried him greatly day and night—had more pain before passing water, half an hour afterwards extending to the groin along the course of the spermatic cord which was enlarged. (6) He had a very dejected look, constant depression of spirits, occasionally a suicidal tendency. (7) Enlargement of the spermatic vessels for which he had worn a truss for several years. (8) His disease had been variously called "disease of the bladder and prostate," "chronic inflammation of the prostate," "irritable bladder," &c., &c. On examination were found two ounces of residual urine common in most cases of prostatic disease and irritable bladder. The urine was turbid but retained a slightly acid reaction and was slightly albuminous. Under the microscope were seen blood globules; specific gravity of the urine was 1,018; it was highly phosphatic, turned thick on heating and effervesced briskly on the addition of nitric acid but on account of the albumen it contained continued cloudy.—Medicine and galvanic treatment were tried.

On the 4th of November he reported himself better: "Very much better; I only pass water three times a day. I have a natural desire to do so, and satisfaction after the act is over."

November 20th.—Without any explainable reason thinks himself worse—irritation of the bladder returned. He states, " I am a great deal more troubled during the night. I lost the nightly discomfort soon after commencing your treatment but the inconvenience during the day continued." Drew by catheter (after he had urinated naturally) two ounces of urine ; it was very slightly turbid through an excess of phosphates. Heat and nitric acid quite cleared it, so that there was now no evidence of blood— could not account for his suffering—altered his treatment and he soon became much better.

January 19th, 1876.—The patient was as on a former occasion very despondent. It appeared that he was desirous of getting married, but feared he would never be able to do so. Further appropriate remedies were employed and towards the end of March he reported himself "quite well." The writer saw him from time to time and found his prostate was reduced to its normal size. He could empty his bladder thoroughly, and the urinary secretion was quite normal. He was accordingly pronounced cured and advised to call if discomfort of any kind should return. But no intelligence was received from him till the following was received—

"October 28th, 1878.

" DEAR SIR,—I daresay you have long considered me ungrateful for not writing as I promised to bear testimony to the beneficial results of your treatment.

" When I consulted you it was the last resource left for me—I never expected to get well as all my previous efforts were fruitless. I was so truly wretched that words cannot express my condition. You know how stubborn my disease was and how frequently I told you that I would leave off all attempts to get well ; how I came week after week and month after month in utter despair. I assure you sir that but for your kind assurance to the contrary and your conviction that I should ultimately be cured I know not what would have become of me. Should you have any patient in the same condition as I was in, please give them the same assurance as you did me and tell them the number of times I came to you shaking my head and saying "no better, no better sir," still thanks to God and you I am perfectly restored to health.

"I am your ever grateful patient,

"W P."

This case was intended for publication in the last edition, but was omitted. The writer wrote to Mr. W. P. in February, 1880, and received the following reply from which it seems that the bladder inconvenience dates much earlier than is specified in the history given of his suffering. In a letter dated 15th February, 1880, the following passages occur :—

"DEAR DR. JONES,—In answer to your note of the 7th instant, it gives me great pleasure to again bear testimony to the beneficial results of your treatment. I think if I remember rightly my bladder first became affected when about eleven years of age, although many years ago I can recollect being continually obliged to pass water. I can also recollect being chaffed by members of my family accordingly. The discomfort left me for a time but returned when I was about fifteen years of age (my present age being twenty-six) and I continued in that miserable state (suffering at the same time from another disease) until I came to see you. I tried one medical man after another but gradually got worse. My life proved a burden instead of a pleasure, when with friends I was unsociable and was so irritable that it seemed I could not treat any one with even common civility. You can willingly show your patients these roughly written lines and if their disease is like mine, bid them to live in hopes, and persevere under your treatment as I did. I need not say more except to again thank you for the attention you paid to me and the care with which you studied my disease while under your treatment. Thanks to your inquiries I have continued well since I last saw you. I shall always be too happy to testify to the benefit I have received at your hands if at any time you feel disposed to refer to me ; trusting you will accept my heartfelt thanks.

"I remain yours gratefully,

"W. P."

No. 11.

Disease of the prostrate of long standing—occupying nearly a year before treatment could be left off—the remote cause gonorrhœa in early life.

B. H., aged 64.

The description given by this patient of the sufferings which preceded his more recent and severe attacks, made it evident to the writer that the gonorrhœa contracted on two distinct occasions in former years, had greatly contributed to his present condition. Though attended by eminent medical men from time to time, gonorrhœal prostatitis had unquestionably existed for some time. In the early part of 1880 he became worse and consulted various medical men. Though fluctuating in his condition under treatment he was nevertheless seldom free from discomfort.

In the year 1881 he presented himself to the writer at the Home Hospital in Dean Street, Soho, with well marked symptoms of chronic enlargement of the prostate. He had frequent urination at night, attended with considerable pain. The urine was cloudy and loaded with mucus.

Soon after the treatment had commenced, under the care of one of the writer's assistants, he had a sharp attack of orchitis, and the treatment had to be suspended. The urinary trouble now became more alarming, and he had to pass water at times every five or ten minutes at night,—the frequency during the day being about once every hour.

On the 27th April, 1881, he came under the writer's personal attention. His symptoms were then about the same as those above described.—Specific remedies and the spray treatment were administered.

The urine, which could not be expelled by natural efforts, amounted only to two-and-a-half ounces, and it took a long time to reduce it, and even when there was evidence of improvement it did not last, as in the majority of similar cases treated by the writer. For instance, the entry in the " case book" under date 18th June, 1881, showed the residuary urine reduced to six drachms only : he nevertheless stated : "I am very much better— the pain is quite gone, and instead of passing water every few minutes I can go two hours, and when the call comes on me while serving a customer I can wait for ten minutes without the same uneasiness I used to experience, and when afterwards I empty the bladder I do so naturally and with very slight pain."

The treatment was pursued and on the 29th of June the residuary urine is found to be seven drachms, *i.e.,* one drachm *more* than on the former occasion.

On the 16th July the residuary urine is reduced to two drachms.

On the 30th of the same month the patient reports himself " all but well."

In August (23rd) he is again a little worse, but the residuary urine amounted to two drachms only.

On the 10th September he is much better again—the residuary urine is only thirty drops.

From this latter date down to the present time he has had little or no discomfort, the amount of urine which he cannot expel from the bladder has not increased and his condition is such as to enable him to say :—" I would never grumble if I continued always as I am now."

NOTE.

This case is by no means typical of *ordinary* chronic enlargement —but of chronic enlargement coming on very gradually after disease contracted in early life. In other words, it was a case of specific chronic inflammation of the prostate. The gland had been enlarging very gradually. The length of time it had been growing had produced more compactness, more hardness so to speak in the gland structure. The enlargement was not great so that the residuary urine never amounted to more than two-and-a-half ounces. The median portion of the prostate was the part chiefly affected— a slight enlargement of which portion will occasion more discomfort than considerable enlargement of the lateral portions, inasmuch as it interferes more with the function of urination. Cases of prostatic enlargement which take a long time to become hypertrophied (enlarged) are always more difficult to cure. And if in addition there is or has been specific disease such as gonorrhœa or syphilis, the cure is still more tedious. Tubercular and cancerous depositions partake of the same character.

These remarks are added to show the reader that much more time is required for the cure of specific cases than for the cure of the more common forms of enlarged prostate which generally yield readily to the writer's treatment.

———

No. 12.

Stricture—cure without cutting.

J. A., aged 48, unmarried.

This patient's sufferings extended over a period of 22 years originating in an attack of gonorrhœa ; the ordinary treatment (by stimulating diuretics copaiba and cubebs, &c.) had been applied. The acute stage passed, a gleet remained, for which he was treated by injections of nitrate of silver and other strong astringents such as lead and copper. Soon after using these, acute orchitis followed and laid him up for several weeks. The inflammatory symptoms subsiding the discharge returned and he became as bad as ever. St. Bartholomew's, Guy's, and Charing Cross Hospitals were then resorted to but with little or no benefit. The patient's own statement further explains consultations from time to time with "private doctors" (as he termed them) without any real benefit but with varying consequences, involving a large amount of continued suffering—fresh symptons following remedies applied—skin disease, serious disturbance and derangement of the digestive organs and general ill-health. Discouraged and despairing (as he said) he resolved to leave "things to nature," and he suffered on more or less for many years, during which time serious symptoms in connection with the bladder were presenting themselves. To use his own words, " he was worried with discharge and difficulty in passing water. The stream became twisted, and sometimes it would splutter in all directions as water coming out of a watering pot." Eventually "he could pass no stream at all—there was perfect obstruction and the urine dribbled away involuntarily night and day—had to wear flannel and other contrivances to absorb the urine as it dribbled away." Though shuddering to think of consulting another doctor, his sufferings nevertheless led him to call in a neighbouring practitioner, who he says, " treated him very roughly while attempting to pass his instruments and made him bleed considerably." At this juncture he was recommended by a friend to send for the writer as one who "never used violence." Accordingly the writer was sent for. The patient was in a deplorable condition when first seen. Most tender treatment was essential. The smallest French bougie could not be introduced without occasioning rigors followed by fever. The flow of urine became less and less and ultimately stopped entirely, and the bladder became greatly distended. To give temporary relief, the bladder had to be punctured and aspirated three times, and in the interval attempts were made to dilate the urethra. A very fine french bougie was ultimately introduced and retained in the canal. At the end of ten days, some progress had been made. The parts were beginning to resume lost function. As the bougies increased in size, the meatus (the narrowest part of the canal) had to be freely divided. Long continued treatment was called for and gradually pursued till No. 38 french bougie was reached. [No. 1 french is about the size of a small bristle, No. 38 about the circumference of a lady's finger (1 in. $\frac{6}{8}$). This will give the reader an idea of the amount of dilatation carried on.] As the patient had intimated his intention to go to Australia, and objected to other or more rapid surgical procedure, it was considered desirable to carry on the treatment of dilatation (thus far successful) much longer, in order to effect anything like a permanent cure, and to obviate retraction of the urethral tissue, which is very liable to occur after treatment by ordinary bougies. The treatment of J. A.'s case was in fact continued for fully three months—a period by no means too extended in cases where permanent benefit is to be gained. As the patient underwent the treatment it was gratifying week by week to see the urine becoming clear, the pain subsiding, the stream increasing in size, and being expelled naturally with the usual force, to which he had been unaccustomed for many years. With this physical improvement his general health also improved—the wretched-looking pinched and pale face, so characteristic of long suffering, became placid and healthful in appearance ; he gained flesh, weight, bouyancy, and courage. The case which gave little or no hope of recovery, responded very satisfactorily to the treatment, although (as already stated) it occupied considerable time. After visiting some friends in the country for a few weeks he made preparations for carrying out his long cherished wish to go to Australia. The friend who had

recommended J. A. informed the writer four years aftewrards that
J. A. was quite well and had not experienced any return of
former discomfort.

NOTE.

The writer has treated several cases of organic stricture in a
similar manner, and has every reason to be satisfied with the results.
The method adopted and applied by him may, to some,
seem tedious compared with other methods, but where a patient
has great aversion to the use of purely surgical means, the writer
would unhesitatingly and with confidence pursue it in preference
to the usual mode of treatment. Stricture treated as by OTIS's
method is by far the most satisfactory, but the writer is acquainted
with patients so treated by other surgeons (one in his own practice)
where contraction showed itself in less than twelve months. The
speedy return of the stricture in that case was very exceptional,
the patient having undergone three operations by different surgeons. The writer is persuaded that where this happens it is due
to want of more assiduous attention, the strictured part is not
divided in its whole length, some fibres of the diseased tissue
being left undivided, and the after treatment by bougies is not
carried out for a sufficient length of time. Ten to fourteen days
as mentioned by Dr OTIS is not sufficient in the writer's opinion
to ensure a perfect cure, and in a manner so vital to the patient's
interest two or three weeks treatment more or less should not be a
matter of consideration to him. For the reader's information
whether professional or lay the writer would add a few general
remarks respecting his own mode of treatment. He has termed
the treatment " unusual dilation," and he is satisfied that unless
this is carried out, that is to say, unless the canal is *unusually*
dilated, the urethral tissue is sure to contract and the stricture
resume its original hard and unyielding condition and the treatment
prove not to be *curative*, but like the ordinary bougie treatment
only *palliative*. One difficulty however is that sometimes patients

will *not* go on with the treatment, saying, " I am quite well doctor ; why treat me any longer ? " and other such-like expressions. The writer's experience however convinces him that the dilatation must be carried on until the area of the canal is dilated beyond the size given by OTIS as the normal size. This appears to have the effect of destroying the tonicity of the structures so to speak and thereby obviating the tendency to retraction as happens after ordinary treatment. Where patients have a dread of cutting as sometimes is the case and there be ample time to devote to " unusual dilation," the writer can confidently recommend it.

No. 13.

Stone in the bladder, weighing one ounce (minus 5 grains).

Mr. G. H., Southampton, aged 63.

About seven years ago while apparently in good health he was suddenly seized during the night with very severe pains in the back extending to the hip of the right side, front part of the right half of the abdomen, the generative organs, as well as the front part of the thigh. This was attended with cold sweats and coldness all over the body. His medical attendant administered the usual remedies, aided by hot bottles to the feet, sides and perineum. He soon got better and was able to resume business again. This attack was clearly that of passing stone from the right kidney. Previously to this attack as well as afterwards he had noticed that the chamber utensil was frequently covered with gravel, but as this gave him no trouble he thought no more about it.

Six months after the first attack he was again seized with similar symptoms and in like manner got better by a repetition of the treatment. Twelve months later however he was seized for the third time.

H2

Soon after this he was quietly sitting in his office when an urgent desire to urinate came on without any warning. In the act of doing so a small stone "shot out" with some violence, which satisfied him he was suffering from stone in the bladder. He suffered little or no inconvenience for a long time after this, but always noticed that he urinated more frequently than other people, which made him fidgetty. As a small stone had passed he flattered himself that he had got rid of his enemy, but such was not the case. He was able however to attend to his business in the usual manner until a year ago when he went out some distance on horseback. The riding exercise increased his suffering tenfold, —the bladder irritation became unbearable, and he passed a large quantity of blood at variable intervals. From this date he became worse until he consulted the writer on the 22nd October, 1879. He was ordered to take rest, and appropriate remedies were administered. When his symptoms had subsided and the writer's spray treatment had got the bladder into a healthier state the patient was operated upon.

The patient was placed under the influence of ether, and a very hard lithic acid calculus was most effectually reduced into fragments by BIGELOW's large lithotrite, and removed by the tubes and aspirator in 55 minutes, including the time occupied in overcoming the impaction of the lithotrites and cleaning them. The *débris* when dry weighed 475 grains, *i.e.*, five grains under an ounce. The only difficulty that presented itself was the impaction of the lithotrites in the narrow part of the urethral canal. The *débris* became clogged between the blades of the instruments. Mr. MELTZER (of the firm of MESSRS. MAYER & MELTZER) who was present saw the difficulty which the writer had to encounter and has since manufactured for him, under his instructions, an improvement upon BIGELOW's instrument which is mentioned in Part IV. The patient made a rapid recovery and returned home in a few days. He might indeed so far as his health was concerned, have left some days sooner.

So completely did BIGELOW'S aspirator wash out the *débris* that not a grain was passed with the urine afterwards. With the exception of slight bleeding from the urethra not a single bad symptom occurred.

Mr. H. called on the writer in January, 1880, and reported himself in perfect health.

No. 14.

Two lithic acid stones, weighing collectively 220 grains.

G. G., aged 67, Leeds.

This gentleman's case presented the usual signs of stone without any complications. The writer, when first consulted by this patient, could not help remarking to himself, "This gentleman *ought* to have stone." He had the appearance of a perfectly healthy man, whom (so to speak) no one would pity—tall, stout, commanding, with a ruddy countenence—the picture of a good-tempered English gentleman. His countenance brightened with a smile when spoken to. He looked less like a patient and much less serious than his anxious son who accompanied him.

Many stone cases occur in such persons. He gave the history of his case in a manner the writer has often wished other patients would follow—clearly and truthfully. No case of stone could be detailed with more accuracy and faithfulness. He said, "I have been suffering, doctor, for two years. I first noticed bits of gravel in my water, attended with uneasiness at the tip of my penis, and a worrying desire to pass water more frequently than natural, which gradually got worse till I had to urinate every ten minutes. I was always worse after exertion, such as riding in a trap over roughish ground. This brought on shivering fits, and not infrequently large particles of stone came away from me with a lot of blood. The pain and uneasiness were always worse after I had passed water, very bad indeed, but went off in a few minutes after considerable straining, until I wanted to pass water again. I am quite well when lying in bed or on the sofa, but as soon as I get up I have to be at it again."

The process of "sounding" soon revealed that he had two small calculi. He was sent to Bolton House, and after slight preparation the stones were removed, under the influence of ether, in nineteen minutes. He made a rapid recovery ; the only little trouble he suffered from afterwards was inflammation of the testicle, which occasionlly—very occasionally—follows the use of instruments in the urethra, independently of stone. The patient soon recovered from his secondary discomfort, and has enjoyed good health ever since.

––––––

No. 15.

Lithic acid stone, half a drachm under an ounce in weight— complicated with two strictures.

G. F. V., aged 38, Widower.

This patient had been engaged in very heavy work, necessitating a considerable amount of stooping. For many years he had been subject to pain in the back, which he generally attributed to his occupation. His discomfort increased, which induced him to employ various domestic remedies, but without benefit. His doctor told him he was suffering from lumbago, and treated him accordingly. Getting no relief he went to St. George's Hospital as out-patient and remained under treatment for three months. At that time he suffered considerable uneasiness, and passed blood in his urine occasionally. Eventually his urinary symptoms became more troublesome, and his urine occasionally became tinged with blood. This led his hospital surgeon to sound him for stone, but none was found. Getting no relief after three months' treatment he went from doctor to doctor without deriving relief until ultimately he left off doctoring altogether and was just as well without medicine. Afterwards additional symptoms troubled him. He passed water more frequently and was in more pain. He consulted another doctor who likewise " sounded " him for stone, but he also failed to find one. Finding ultimately and after enquiry that the writer had treated such cases with success, he accordingly in September, 1879, presented himself with very severe symptoms. An examination revealed two strictures in the urethral canal, one two inches from the meatus and another about five-and-a-half inches. Both were very contracted strictures admitting only No. 2 French catheter. In the course of a few weeks he was much relieved ; still he passed blood with urine and had considerable straining after urinating. A microscopic examination revealed lithic acid. An examination for stone by " sounding " made his case beyond doubt. He was advised to enter the writer's " Home Hospital for Stone." He was placed under ether, and in five minutes under the hour a calculus weighing an ounce less half a drachm was removed. He made a rapid re-covery and left the Home Hospital in a fortnight. Since then he has been perfectly well—all his painful symptoms (his back-ache of seven years' duration included) have disappeared—and the writer has been informed that he is now married to a second wife.

G. F. V. called on the writer in January, 1882, in perfect health and has been so ever since he left the Home Hospital.

No. 16.

A remarkable case of stone, complicated with a very large prostate, occupying the writer three months before he could get entrance into the bladder and " sound " the patient.

J. P.

J. P. had been suffering from urinary trouble for a long time, and had been getting gradually worse, but more particularly during the last eighteen months.

Getting no relief from the treatment of several medical gentlemen he had consulted (in his own neighbourhood) he came to a London physician associated with Guy's Hospital, whose prescriptions he followed but without benefit. He now sought the writer's advice. His symptoms were as follows:—

(1) Intense irritation in passing water, which he has to do every hour or two hours during the day but less frequently at night. (2) He is always worse for two or three minutes after emptying the bladder. (3) The pain extends to the perineum (crutch) and fundament, so that during each act of urination there is severe involuntary straining in the rectum (back passage). [The above symptoms were so severe that the writer entertained serious doubt as to whether there was not some malignant disease of the lower bowel. The spasmodic seizures were so violent, sudden, and uncontrollable that the patient could seldom reach the commode without great unpleasantness.] (4) At the commencement of his illness he passed a large quantity of blood after emptying the bladder and after any kind of exertion. (5) He has been examined for stone four or five times by his various medical attendants, who assured him one and all he had *no* stone. His London physician however told him he "thought he had." (6) Is *worse* when lying on his back, and easier when lying on either side, and easier on the left than on the right. (7) The urinary secretion is loaded with mucus, pus, and blood. (8) An examination *per anum* (by the lower bowel) discovers the *largest* prostate gland the writer ever examined. A small elastic gum catheter was introduced with some difficulty, and six ounces of very fœtid residuary urine, mixed with blood and mucus, was withdrawn from the bladder. (9) Under the microscope are seen pus and blood globules and crystals of oxalate of lime, "dumb-bell" as they are called (from their resemblance to a dumb-bell). (10) The prostate so large that an ordinary "sound" could not be introduced—the passage also so sensitive that the operation of "sounding" was abandoned for another day. He was,

moreover, very corpulent, and it was with the greatest possible difficulty he could remain on his back without a sense of painful suffocation—in fact, it was postponed till it could be properly done under ether. In about a week this was tried without effect. He took the anæsthetic very badly, and it had to be left off on several occasions to avoid accident. After repeated attempts it was found on examination that the part of the urethra corresponding to the prostate was dilated into a pouch, which resembled a second bladder, so that the "sound" *appeared* to enter the bladder but did *not* do so. He was examined over and over again by the "sound," but it never could be coaxed to enter the bladder ; this will account for the patient's former medical attendants failing to find stone. These examinations tried the poor fellow sorely. The symptoms above described appeared very conclusive, and indicated that he had the worst and hardest form of calculus, *oxalate of lime*, complicated with an enormous prostate. Still, it is a rule in surgery never to assert the presence of stone without the *physical* proof which "sounding" elicits.*
To afford permanent relief it was clear the foreign body (stone) must be removed, and it was all but decided at one time that the operation of "cutting" (lithotomy) was the only means of adopting to relieve him, but considering how fatal the "cutting" operation (lithotomy) has been, and considering also the peculiar constitution the writer here had to deal with, a resolute attempt was made to dilate the urethral passage and reduce the size of the prostate and remove the stone by the writer's method aided by BIGELOW's operation.†

It was clear that the first thing to do in this case was to attempt to do what high authorities say *cannot* be done "by any known

*Cases are on record where patients have been "cut" for stone and no stone as been found. Cases have also been recorded where the stone has been struck (sounded) with the instrument called "a sound" at one time and not at another.

†One in about three-and-a-half to four patients treated by the "cutting" operation in adults die.

means,"* but which the writer maintains not only *can* be done but *has* been done, viz., to *reduce* the size of the prostate. The writer's system (the spray treatment) was now put to a fair test, and as this patient (and scores more) can testify he was greatly relieved during the spray treatment.

In this case the enlargement was in the *median* as well as lateral lobes—the quantity of residual urine (only six ounces) clearly indicated that such was the case. A very small amount of enlargement of the *median* lobe gives a considerable amount of discomfort, while a very great amount of enlargement of the lateral lobes might exist without much discomfort being co-incident with the disease. After treatment for the prostate had been employed for some time it was thought advisable to employ additional means with a view of enlarging the urethral canal for the purpose of " sounding." This was done for some time, and another attempt was made to "sound " the patient again while under the influence of ether. The ordinary "sound " would not pass into the bladder, notwithstanding repeated attempts by the writer aided by attempts made by three other medical friends present. *All* attemps were ineffectual—it was found impossible to " sound" him. The question of " cutting " was put before the patient in a day or two as being the only means of getting rid of the stone. The patient replied " but, sir, you can't be certain, you say, there is stone without sounding me, which appears to be the only safe test." The writer could not but assent to this as true, and he might have told the patient that patients had been (apparently) sounded and afterwards cut for stone and no stone found. The rule is that the "sound " must *touch* the stone by more than one surgeon *at the time* the operation is about to be performed. In this case this could not be done, the "sound " would not go into the bladder. All these circumstances weighed heavily on the patient's mind as well as on that of the writer. The

---

*See opinions of Sir HENRY THOMPSON, VAN BUREN, KEYS, GANT, and others in article on the prostate (pages 54, 55, and 56 in this publication).

writer was *morally* certain (though not physically so) that stone was present, still he *might* be mistaken in his diagnosis, and a fellow-being's life might be sacrificed. The treatment was proceeded with and with good results.

Finding the ordinary (stiff) "sound" could not be introduced into the bladder, in the emergency the writer attempted another kind of "sound." The ordinary "sound" appeared to be obstructed long before it reached the bladder, which he thought was due to its being straight and stiff. On this account he asked Mr. MELTZER of the firm of MAYER & MELTZER to make a longer "sound" with a flexible shaft and metallic tip. A shaft of this kind of "sound" would give or bend according to the tortuosity of the canal. The circumstance that actuated the writer was this: If (he argued) there is a stone it would no doubt be found behind the enlarged prostate. Even if a stiff "sound" could be coaxed into the bladder the stone would not be reached, consequently the physical test (the most reliable one of all) would not be available. He argued also that if a flexible "sound" could be coaxed into the bladder, it might be cautiously glided along the posterior part of the bladder, and meeting with an obstacle the flexible handle would bend and bring the metallic tip forward towards where the stone usually lies. The writer's view proved true—the first trial verified his reasoning—the stone was at once discovered. The stiff "sound" could never have been brought in contact with the foreign body. It is not to be wondered then that J. P.'s former medical attendants failed to discover the stone, but the writer is of opinion that at the stage of the disease when he was examined the "sound" never really passed beyond the pouch-like dilatation already described. The wished-for object now accomplished, the writer was encouraged, as was the patient, and it was decided to continue the spray treatment for awhile longer. "If I can be cured (said the patient) without the knife, I won't mind the time occupied. I am wonderfully better already." He also said, "They are grumbling at home about the length of time I have been away, and many have said that I should never return home alive, but by God's blessing I hope to disappoint them." The treatment was continued with vigour, and by the aid of dilators and the spray treatment a lithotrite was ultimately got into the bladder. Accurate dimensions of the stone, said not to exist, were soon taken, and in a few more days a hard stone, composed of oxalate of lime was removed, and the suffering patient was speedily relieved.

Mr. J. P. visited London in November, 1881, in perfect health and well pleased with all that had been done for him.

No. 17.

Oxalate of lime stone, weighing nearly four ounces.—Two instruments broken during the operation.—Recovery.

G. S., aged 28, married.

The following narrative of this patient's sufferings was given by himself, and it is here presented *in extenso*, partly as information to other like sufferers, but chiefly as both suggesting and justifying the remarks appended respecting the instruments used in this particular case. The patient's statement is as follows :—At the early age of 19 years he had a most unusually constant desire to urinate, accompanied by pain at the end of the penis. He kept on at his work (fireman on a locomotive engine) for twelve months, when he got worse, passed blood, and had to urinate every few minutes. He consulted a doctor at Castle Bridge, who sounded him and suspected stone but could find none, and recommended him to a doctor at Carlisle, who also sounded him, discovered a stone, said it was a large one, and sent him into the Carlisle Infirmary, of which he was a consulting surgeon. He went in, and the house surgeon and some of the visiting surgeons examined him several times, but could find no stone. After he had been an in-patient for six weeks and had been examined several times

unsuccessfully, the before-mentioned consulting surgeon came in
and said, " Why are you keeping this patient so long ? " and he
was told that the patient had not got a stone. Upon this he sent
for his instruments, and using them at once discovered a stone.
Two weeks after this he was put under chloroform and
again examined by the consulting surgeon who said the stone
was smaller than on his first examination he thought it was, and
that it was prickly like a horse chestnut. It was decided that the
patient should be " cut," and he was told that the " cutting
operation " would be better, and not so painful or protracted as the
crushing operation. He had now been in the hospital nine
weeks; and at the end of eleven weeks he had to leave the
infirmary owing to domestic troubles which culminated in the
death of his wife. His father now would not allow him to return
to the infirmary as he was afraid that the operation might prove
fatal, so the patient started work again. Soon however he began
to pass blood in the urine, and a friend advised him to try a
medicine known as " Dutch drops." This stopped the blood and
brought away a great quantity of mucus, and in fact did him much
good, but it soon lost its effects and he got as bad as ever. He
was then off work for eight months, when although still very ill,
he tried to start work again. He fortunately got an easy place
and worked on and off till Christmas, a period of four months,
when a friend recommended him to try some herbs which he
declared had dissolved a stone from which he had been
suffering for a long time. He accordingly took these herbs for a
week but got much worse and gave them up in disgust. He
began work again, and in about another week he thought the
stone was being dissolved and passing away as he passed a
quantity of urine of a sandy consistency. From this time he
gained flesh and rapidly got better. He worked at his previous
occupation, and continued to do so without any very material
inconvenience for five years. It should be noted that although
apparently well (for he concluded that the stone was entirely

dispersed), his urine was always more or less clouded during the whole of these five years.

The patient had in the meantime married again. About two years ago when on his engine he suddenly felt something move in his bladder, accompanied by the old pain at the end of the penis. This discomfort lasted for a fortnight, and he then got better. He continued to work for another twelve months, during which time he was sometimes better and sometimes worse. The following Christmas however he broke down entirely. All the old and worst symptoms came on again. His urine dribbled away from him involuntarily, and he continued in this state for three months. He then got slightly better again and went to work for seven . weeks, but ultimately soon succumbed to a return of his old symptoms—at this time he lost all control over his bladder.

The foregoing statement narrates the circumstance under which this patient consulted the writer, and which led to arrangements being made to receiving him into the Home Hospital in Dean Street, Soho, on the 12th day of September, 1881. He was examined on the same day by Dr. JONES, who at once found a very large stone and determined to operate. Unfortunately the patient caught cold, and the operation had to be postponed for nearly a month. After preparatory treatment to subdue the irritable and inflammatory state of the bladder, which the spray treatment most effectually does in some cases, the patient was placed under the influence of ether. The stone was seized with one of BIGELOW's large instruments, but was soon found to be so large as to prevent the lithotrite locking, and though seized in several directions the screw would not bite. An attempt was then made to bring the male and female blades together, by forcibly opening and closing the blades, but the stone would not give way. The writer now attempted to break the stone by tapping with justifiable force with the ball of his hand. This was repeatedly done without effect.

The stone measured two-and-a-half inches in the largest direction, two-and-a-quarter in another, and two inches in the narrowest direction, thus showing it to be nearly as broad as long, —in fact, nearly alike in measurement in all directions. Until BIGELOW's lithotrites were introduced, surgeons never attempted to crush an oxalate of lime stone larger than a small chestnut, but unhesitatingly reserved it for the "cutting" operation. Authors have written of this kind of calculus—"The lithotrite recoils from it and feels as if a bit of iron were between the blades." What was to be done? lithotomy (the "cutting" operation) seemed the only practical way of removing the stone. Knowing, however, the fatal results so often attending this operation, as applied to large calculi, in cases other than those where the patients are children or under the age of 21, the writer resolved to try the effects of comminuting the stone with a hammer, as had been successfully done in the case of E. H. (see cases). Every precaution was taken to avoid injury to the bladder by sharp fragments which might unavoidably happen in forcibly crushing such a stone as he had to deal with. For instance, the bladder was filled with warm water—the hard stone was brought as near as possible (between the blades of the instrument) to the centre of the bladder before the process of crushing commenced, thus securing an arrest, so to speak, in all directions, of the splintered fragments—in other words, the fragments must in this way be impeded in all directions before reaching the walls of the bladder. By this means mechanical injury (the chief drawback to ordinary lithotrity) was avoided.*

---

*For many years in the performance of ordinary lithotrity, the writer employed with considerable advantage bland and thick, yet soothing fluids, such as gruel, linseed tea, gum water, quince, glycerine, and emulsions of various kinds. These fluids, he is assured, are of great use, since they intercept to a great extent the contact of hard fragments with the coats of the bladder, and by this precaution mechanical injury to the already inflamed bladder was done away with. With his own mode of treatment, however, this is not necessary, inasmuch as his preparatory spray treatment in most cases subdues inflammation. Nevertheless, he still, in some cases of very irritable bladders, employs these bland and soothing fluids, and in case of lithotrity, or even BIGELOW's operation, in the hands of other surgeons, he would still strongly recommend that plan of treatment.

All things being arranged, and the stone being well secured between the male and female blades of the lithotrite, the male portion was struck with a good sized hammer severely and successively, when, after a considerable amount of hammering, the stone gave way. On attempting now to seize again other fragments, the lithotrite was found to be crippled. It had to be withdrawn and substituted by another of equal strength. A large fragment was now seized measuring an inch and a half. All the strength the writer could bring to bear was exerted. The stone, now reduced to a much smaller size, was still too tough. Attempt after attempt was made, but without effect. Dr. BUCK, a medical gentleman present, was asked to steady the handle of the lithotrite while the writer was working the screw. Finding his attempts were ineffectual, another gentleman present was asked to try his strength, but failed in his attempt. The physician who had charge of the anæsthetic (ether), a powerful man nearly six feet high, and proportionately strong and muscular, was asked to try his strength. He also failed. The hammer was again employed, but this lithotrite, like its predecessor, came to grief. An attempt was made to unlock it, but it was found impossible—it was immovable. Fears were entertained that one of the blades of the lithotrite within the bladder was bent. It was, however, soon found that the obstruction was in the handle. It was found that the concussion of hammering had, by accident, partly locked the instrument. The male blade of the lithotrite had been battered (so to speak) into the female blade and the instrument became perfectly disabled. A messenger was dispatched to the makers of the instruments asking them to bring proper tools to overcome the difficulty. Messrs. MAYER & MELTZER were soon on the spot. The instrument was unscrewed and liberated, the obstruction being (as anticipated) found to be in the handle. The operation now proceeded, large instruments being employed, and in the course of one hour and thirty-five minutes (including the delay) the whole of the stone was satisfactorily removed with the exception of one small fragment, which was removed in about a week after the operation. The patient made an excellent recovery, and has continued perfectly well ever since.

NOTE.

The above case suggests a few remarks, additional to those already made in the earlier part of this present edition, in reference to the objection urged in various quarters against the use of large instruments. His own experience convinces the writer that down to the present time lithotrites including those used by Professor BIGELOW, are rather too *small* than too large.

The above case is in proof of the writer's view. One reason urged in favour of small lithotrites is that we now discover calculi before they reach the magnitude of former years where they were occasionally found to nearly fill the cavity of the bladder. This is no doubt true in many instances, but cases of very large calculi still every now and then present themselves. The above is only one of several instances coming under the writer's notice,—cases in which "small and handy instruments" would not, could not be of effective service. Even where a calculus does exist, circumstances sometimes afford the patient temporary relief, nevertheless the stone is in the bladder and grows insensibly larger and larger, without it may be exciting much attention. It was so in the above case, it was so in J. B.'s and E. B.'s case, and it was so in the case cited by Mr. CADGE, of Norwich (see page 7).

The whole history of oxalate of lime calculi, as suitable in the one case for crushing and in the other for cutting confirms the conclusion that the writer has ventured to express. With the view of meeting emergencies unexpectedly arising, the writer has had an instrument made specially for his use by Messrs. MAYER and MELTZER,—an instrument which supplies the deficiencies found even in BIGELOW's instrument. In the first place the instrument thus specially made is stronger in every

I

direction. Next, BIGELOW's instrument has nothing on the handle to indicate whether it is locked or open. This accounts for the accident unexpectedly occurring in the above case. The reader will remember that when it was found that the lithotrite did not lock on the stone, a hammer was used which doubtless broke the stone into several fragments, and the instrument by the jar occasioned became accidentally partially locked, and continued hammering led to the disablement of the instrument, in which state the lithotrite could neither be locked nor unlocked until it was taken to pieces—the male screw was jammed into the female screw. Again, even if BIGELOW's instrument were strong enough, its globular handle does not afford sufficient leverage for crushing a stone like that in the above case, and the palm of the hand after usage in a prolonged case becomes blistered and so far disabled. The writer has therefore had fixed to his new instrument a long transverse handle seven inches in length, affording considerable leverage, and, in case of necessity, two persons can work it, while the larger handle and screw respond to the increased power exerted, and should it become necessary to use the hammering process the index, as well as the long transverse handle, will at once indicate to the operator its locked or unlocked condition, and here again a defect found in the ordinary lithotrite is obviated.

The reader must not be alarmed at this array of mechanical force. The force is *not* directed against the urinary bladder, or indeed against any part of the urinary apparatus, but only against or upon the *stone*, and ought (in dexterous hands) hardly to come in contact with the bladder at all. The instrument is *within* the bladder, in which there is abundance of fluid to keep the walls of the organ apart. The writer employs, or has at hand, in these cases a large and powerful mechanical appliance, just as he would employ a powerful horse for hunting or carriage work. A horse of more power than may possibly be requisite for the weight, does not injure the rider or carriage if properly handled, but there is reserve power at hand, and available, should necessity for its use arise. A large instrument may as readily dispose of a small stone as of a large one, but a small instrument cannot dispose of a large stone, hence the larger, and more powerful instrument is on all accounts to be preferred.

The writer conscientiously believes that with such appliances as above described, the " cutting " operation in adults might be abandoned, save, perhaps, a few very exceptional cases. In a few more years he hopes to be enabled to offer an opinion from accumulated experience in his own practice, and that of others— showing the results of BIGELOW's method when dealing with stone in children and persons under the age of 21 years, as compared with lithotomy, which, as applied to children and young people has been very successful, while, as applied to cases in advancing and advanced life, it has proved not only dangerous, but very fatal.

No. 18.

Disease of the bladder of long standing, with dangerous condition of the kidneys, and other complications caused by stone which had been overlooked.

[This case, as well as the preceding one, is presented as showing the importance of having at hand, and available for use, me- chanical appliances of sufficient power to meet emergencies arising where the stone is found to be unusually large and hard.]

E. T., aged 39, married,

Had he stated been suffering for thirty years. When only nine years old he had pain in the back, and blood was noticed in his urine in considerable quantities on several occasions. His mother was told that he had disease of the kidneys, and the doctors feared he would not survive the age of 17 or 18. Besides blood, which was said to come from the kidneys, he suffered pain in urinating, usually worse at night, and after emptying his bladder.

His school fellows accused him of being idle, as he seldom joined in school games—running or jumping he could never indulge in, as violent exercise of any kind greatly distressed him. From the age of nine till eighteen he suffered off and on great pain—always wanting to pass water. On several occasions the school-master accused him of " shamming," till his parents assured him and satisfied him it was not so. About the age of eighteen his symptoms became gradually more urgent. His condition fluctuated during several years, and although always looking ill, he got on with his work pretty well. He was told that married life would improve his health, and he accordingly married, but became worse, and had to " declare " on his club for many weeks at a time. The certificates given him by the club doctor stated that he was suffering from "inflammation of the kidneys and bladder." In the early part of 1880 he consulted one of the writer's assistants, and got some relief, but, as hitherto, he was laid up from time to time under the club doctor.

On February 19th, 1881, the writer saw him for the first time, when he said that he had been " on the club for some months, and that the club members were getting tired " of him.

His symptoms were :—(1) Considerable pain night and day, but " worse at night." (2) He had to urinate about every twenty minutes. (3) The urine was intermixed with mucus and blood. (4) The urine, loaded with albumen, was very offensive. (5) His countenance sallow and careworn—eyes puffy. (6) No appetite, (7) Constant vomiting. (8) Breath very short. (9) Ankles and legs slightly swollen. The writer noted in his case book :— " poor fellow, death is depicted in his face."

On examination a large calculus was discovered—introduced catheter—twenty-four ounces of ammoniacal urine withdrawn—washed the bladder with water at 98°, acidulated with acetic acid. These operations over, a spray was contemplated, but the catheter having become impacted by the large stone, an application had to be administered through the catheter. After a little gentle manipulation the catheter was withdrawn, and instructions were given to empty the bladder twice a day and to wash it with a warm solution, which was done. He was seen and treated several times during the next month, giving decided evidence of continued improvement.

April 2nd he said, " I am decidedly improving in every way ; I have less pain. I go an hour-and-a-half to two hours at a time during the day, and at night I am also better in myself. I go half-an-hour and three-quarters of an hour at night, which is always my worst time."

April 23rd.—Told to empty his bladder—result of effort, one ounce and a half, clear and without pain—drew by catheter twenty-six ounces—very little mucus, but urine still albuminous —more pain in his loins than he has had for weeks.

May 7th.—" Slept two hours without awaking last night, and feels better day and night than for many months. The urine, however, comes away unconsciously when asleep."

May 21st.—Has made further progress. Ether had been given three times at intervals from the 7th May, which he took well, and, with the exception of extensive mischief in the kidneys, great progress made in general health. He has still pain over his loins, also headache. He was made aware of the risk of the operation of removing the calculus by cutting, and of an equal risk, indeed, of any operation, owing to his bad state of health, and particularly having regard to the disease in his kidneys. Both he and his friends, who knew the improvement he had made under the preparatory treatment, felt sanguine of success. So, indeed, did the writer and several medical friends who had seen him. The kidney mischief was a chief object of concern, still it was hoped that on the stone being removed, and the health of the bladder restored, that mischief would itself become corrected. It was, therefore, decided to perform the operation on Sunday, May 22nd. The stone was found to be a very hard lithic acid formation, and measured two inches and three quarters in one direction, two inches and an inch and a half in other directions, tapering, therefore, towards one end. Ether was administered. The stone was readily seized, but the lithotrite would not lock. Tapping, however, with the hand soon crushed a portion of the stone, and the remainder of the process was less difficult. Some of the fragments were evidently round and hard, and kept continually slipping away from the blades of the instrument, and for some time a process of nibbling (so to speak) was repeated. When the patient had been under ether for an hour, the gentleman who had charge of it, having never before administered it in a prolonged operation, became very anxious and nervous. The writer seeing this released him from his responibility and administered the ether himself, handing over the lithotrite to another gentleman who was present, who completed the operation in another thirty-five minutes, most satisfactorily.

May 23rd, 9.30 a.m.—Has passed a somewhat restless night, constant micturition, and passed some blood, but on one occasion slept an hour-and-a-half, face pinched, nails blue, hands quite cold, temperature 101.

Same day, 4 p.m.—Temperature risen to 102·2, hands still cold and nails blue, slept an hour.

Same day, 8.20 p.m.—No perceptible change, but is cheerful, and has slept an hour-and-a-half since 4 p.m., urination still painful.

May 24th, 10.30 a.m.—Temperature 101·1, slept three hours during the night and shorter intervals at variable periods; passed, since the operation, blood and mucus in good quantities, but no *débris;* urine clearing, takes food well, is very cheerful, hands quite warm; urine still clearing, and he passes it with more ease.

May 25th.—Improving in every way; countenance natural, slept well at intervals during the night, takes food well, temperature normal, water still cloudy, and the patient says :—" What a comfort it is to be able to pass water so freely and in such quantities !" '

May 28th.—Does not seem quite so well in himself, a little blood again intermixed with the urine ; slight fever. On examination a small fragment of stone discovered in the bladder, but at once easily removed.

May 31st.—Much better in himself, but complains of pain in the loins.

June 5th.—Has continued gradually to improve, but, for some reason, the urine is cloudy and fœtid.

June 10th.—Improving but " thinks he has taken cold, which has increased his back pain ; " the urine has again become cloudy and is very fœtid, feels chilly, and has no relish for solid food. The urine is albuminous and alkaline. He is suffering from secondary cystitis, which usually occurs about the fourth to the seventh day. This condition at so late a period (eighteen days after the operation) is very unusual. The urine is intensely fœtid, resembling the fœtor noticed in cancer of the bladder. The temperature is 99°.

June 22nd.—Has had a bad night, and appears to have had a shivering attack, very much resembling urethral fever. Fear is entertained of suppuration at the seat of pain in the loins ; temperature 100·2 ; cheeks flushed. To have a bath at 100°, and another in twelve hours, if not better. Aconite to be given in five-drop doses of the first decimal dilution every two hours, and a compress of Hepar Sulphuris to be applied to the seat of pain, and the same remedy to be substituted for aconite in twelve hours, if not better.

June 23rd.—Is much improved—no return of fever.

June 24th.—Continues to improve.

June 28th.—Is very much better ; takes food with a relish and wants to go to the seaside for change of air.

From the above date he continued to make fair progress—the variations being due to his sensitiveness to cold, and affecting his kidneys. Although he did not quite recover while in the hospital, he was sufficiently restored to take leave of it, and before going to the seaside he went to his own home for a time. While making preparations to go into the country the kidney mischief again presented itself, his general health failed, he had several severe shivering attacks, and considerable swelling over the region of the right kidney. Soon afterwards he felt something " give way " in that region. The urine became very much thicker and very fœtid, the feet and legs began to swell, and it became evident he could not long survive. An abscess had formed in the kidney, which broke and discharged itself into the bladder. He soon became comatose, and died from a complication of disorders—blood poisoning, bronchitis, and abscess of the kidney, and notwithstanding the very successful removal of a large stone weighing 5 ounces and 30 grains. Although the above patient died, it cannot be said that the death was due to the operation. The length of time he suffered with stone (though not discovered), the irritation and disease of the kidney—in a word, long standing disease—contributed chiefly to the fatal result.

No. 19.

Two small calculi, weighing collectively only twenty grains. Patient unsuccessfully sounded by several surgeons.

F. L., aged 42,

But appeared many years older, stated that " he had been a great sufferer from urinary trouble for six years—had been under the treatment of several surgeons of celebrity without relief, who agreed in saying he had kidney disease." He looked pale and emaciated, and had the appearance of a man suffering from what one of his medical attendants called his disease, viz., "Bright's disease of the kidneys." Finding no relief from the surgeons consulted, he went to St. George's Hospital. He had been examined for stone by all the doctors he had been under, but no stone was found. The writer's record of his case (in addition to the foregoing), is as follows :—" His appearance is pale, his countenance giving evidence of great physical suffering. The cellular tissue under his eyes is slightly swollen and puffy. He says he is worn out with pain and anxiety on account of not being able to attend to his business. He has passed eight or nine small fragments of gravel at uncertain and irregular intervals from the commencement of his suffering. Has always more pain after urinating, and after unusual exertion. Occasionally while in the act of passing water the stream of urine suddenly stops." The writer gave the patient to understand that he had most of the typical evidences of stone, and that if he did not get better he must be again examined for stone, but he replied, "I don't think it is of any use; all the doctors have examined me for stone but none of them found any." He continued under the writer's care without any benefit for a considerable time, and ultimately consented to enter the Home Hospital. He was "sounded" from time to time for stone without success. Eventually he was examined in the standing position, and a small foreign body was discovered. After some preparation he was placed under ether, in the presence of several medical gentlemen. Although the writer was absolutely confident there was a small calculus in the bladder, he searched in vain for more than twenty minutes. The medical gentlemen were asked in their turn to search for the foreign body, which they did, each in turn, and each confidently affirming that the writer had " made a mistake," and that if there had been a stone in the bladder it must have been found. The writer tried once more and brought away a fragment not larger than a barley-corn. The operation was discontinued and the patient put to bed. In a few days the patient said he was better. At the end of a fortnight the urine did not clear, and the discomfort, though not so bad, was still very trying to the patient. A lithotrite was now introduced and another fragment laid hold of and extracted. The patient rapidly recovered, gained health and strength. The urine became quite clear, he lost all his trouble and has been perfectly well since, and continues so now—after the lapse of several years from the time when the small fragments were removed.

It is difficult to understand how such a small body could bring a man nearly to death's door, while, in another case the foreign body (alluded to on page 17) weighing nearly nine ounces remained quiescent and harmless during many years of a man's lifetime.

No. 20.

Stone weighing only eighteen grains. Patient sounded in London, Liverpool, and Manchester unsuccessfully.

J. B., aged 52.

Had been suffering for nearly five years, under several physicians and surgeons of distinction in Liverpool, Manchester, and London, and had been "sounded" for stone about twenty times. When he came to the writer the following was recorded: "Has all the symptoms of stone in a marked manner, has to make water twenty times a day, passes blood with his urine every time he goes to stool, and occasionally at other times; passes water in a moderately full stream, but occasionally it stops suddenly during the act; is worse after his day's work; being a town traveller he has to get in and out of his trap a great many times during the day; has great discomfort at night when in bed, and passes water six or seven times during the night, and after doing so has great discomfort in his rectum, and a feeling as if his bowels wanted to be relieved; has drank gin and whisky pretty freely. The urinary secretion is normal in quantity, and is of an acid reaction, but traces of blood are seen when examined under the microscope; has an ounce and a half of residuary urine."

He was treated by the writer for three months with considerable benefit, and was perfectly satisfied with the progress he was making, until one day he was seized with intense discomfort, and passed, he thought, "about a pint of blood." The writer was summoned to his bedside, and after some trouble the blood was arrested by suitable treatment, including the application of ice into the lower bowel. In about a fortnight he was well enough to be removed to the Home Hospital.

After a number of " soundings " a small hard stone was discovered. He was prepared for the operation of removal, and a day was appointed for its performance. When well under ether a search was made but no calculus could be found. Four medical gentlemen present were asked to search, and each in succession said he could "find no stone." The chloroformist was now asked to search. He did so very carefully, but with the like result. The patient was under ether for a full hour, and during this time the writer was rather taunted—"Ah, doctor! you have made a mistake this time!"—in fact, all present concluded that the writer had made a mistake. The patient was put into bed and further operation was postponed.

Four days afterwards one of the gentlemen who had been present at the unsuccessful operation called at the Home to enquire after the patient and was told that he had had a good deal of hœmorrhage, but that the bleeding had been arrested. He was invited to visit the patient, which he did in company with the writer and one of his assistants. The lithotrite was introduced, and in an instant the stone was seized and brought away bodily on the spot. From that time the patient improved daily and in a few days all the unpleasant symptoms passed off, the urinary fluid gradually cleared, and very shortly afterwards the gratified patient left the Home perfectly cured.

Probably so small a stone could not have been in the bladder during the five years the patient had been suffering. Doubtless the prostate and bladder were first diseased, and gave rise to symptoms of stone (which now and then happens). During treatment before and after the stone was removed, the bladder and prostate had been cured by the spray—at all events, the patient informs the writer that he has had no return of his unpleasant symptoms.

No. 21.

Stone unsuspected—enlarged prostate—suffering protracted, chiefly through not being sounded. Patient assured by two surgeons there was no stone and sounded by another surgeon unsuccessfully. Sounded later on, and two oxalate of lime stones discovered and removed.

E. O., aged 67,

Consulted the writer in September, 1879, after having been treated by several medical men, three of whom were eminent surgeons. It appeared that the patient's urinary trouble had commenced about six years previously to 1879. A discharge was seen in the urine and was called by the medical man then consulted "gravel gout." Medicine was administered for three months, but without relief. Another eminent surgeon next prescribed the use of the catheter at repeated intervals. This being done some relief was experienced, and the patient continued better, more or less, for some time. Not satisfied, however, with the progress, he consulted another practitioner, but still without relief—the discharge and discomfort continued. Another surgeon was now consulted, and by his suggestion quinine (first introduced by Mr. NUNN, surgeon to Middlesex Hospital) was injected, but without avail. When first seen by the writer the left lobe of the prostate was found to be enlarged ; the right lobe also, but not to the same extent. The urinary secretion was loaded with mucus, the deposit representing forty-two per cent., the residuary urine being six to eight ounces. Under the microscope it gave evidence of blood and oxalate of lime crystals. After several applications of the writer's treatment had been used the urine cleared and the mucus was greatly reduced, but the residuary urine was never reduced beyond five ounces.

On several occasions the writer observed to the patient that he distinctly recognised something against the catheter like a stone, and suggested the propriety of "sounding." The patient objected to this, remarking that as he had none of the usual symptoms of stone he would not consent to be "examined for stone." Meanwhile the patient improved under the treatment, and was so satisfied with what was being done that hesitation was felt further to urge examination for stone. And inasmuch as, for a time, the sensation (as of a stone) was not felt, the patient persuaded himself, as others had done, that stone was not present. The treatment having to be prolonged, and the patient not recovering, the writer remarked to him that his case was almost the most obstinate one he had ever treated. Eventually, and as no real improvement showed itself, it was again urged, and agreed to by the patient that "sounding" for stone should be gone through. An examination (under ether) was made, and a very hard oxalate of lime stone was found—also another smaller stone. A day was fixed for removing them. On the 12th day of June, 1881, ether was administered, and the stones were removed in fifty minutes. The time occupied was longer than usual, having regard to the size of the stones (a little over half an ounce), and owing to the enlarged prostate and the hardness of the stone and other complications.*

Since the above operation the patient has been well, but never able to discontinue the use of the catheter, owing to the enlarged prostate which still exists. The writer has not been able, since the removal of the calculi, to apply his *special* treatment for the prostate. The patient has been moving about from place to place with his family, and quite contented with the removal of the stones. Having lost all his serious discomfort, and sleeping well, he is quite satisfied with the improvement effected.

---

* The writer considers it safer to perform all these operations with perfect deliberation. Hurrying, as some operators have done, is neither justifiable nor necessary, since it is proved beyond doubt that if a patient be kept fifteen to thirty minutes longer under ether, it does not in any way protract his recovery, while hurrying over it is likely to injure the coats of the bladder—or, fragments may be left behind, causing the patient great discomfort afterwards.

No. 22.

Stone in the bladder—not discovered by several surgeons.

J. B., aged 54, single.

Gave the history of his case as follows :—" Early in 1875 he began to suffer inconvenience in the bladder. In the summer of the same year he noticed gravel in his urine, attended by frequent desire to urinate, and a burning sensation at the end of the penis. The gravel soon ceased to pass, but the urinary discomfort continued, — he had to urinate at shorter intervals. This continued till the spring of 1879, when the irritation became more troublesome—the urine became cloudy, and tinged with blood and soon after pure blood passed. His condition fluctuated till the end of 1880. He had consulted several surgeons, was under treatment in the North London Hospital for two months, and although several times sounded for stone none was discovered.

On the 29th January, 1881, he consulted the writer. His symptoms then clearly indicated stone ; but, in addition, he had also symptoms of prostatic mischief, and very severe straining had, moreover, caused an inguinal rupture. He urinated almost every hour. The chief pain was at the bulb of the penis and lower part of the belly. He was very irritable and excited, his pulse 140 per minute, temperature nearly 101 ; looked ill and thin. The urine contained blood as well as uric acid crystals. Sounding soon revealed that he had a hard uric acid calculus.

After appropriate preliminary treatment he was, on the 9th February, 1881, operated upon for stone, under the influence of ether. The *débris* removed was composed of lithic acid, and weighed three-quarters of an ounce. He made a good but slow recovery. In a fortnight after the stone was removed treatment was directed to the enlarged prostate. This discomfort gave way slowly, but was not cured till the end of March in the same year. The writer has seen him twice since that time. In January, 1889, he was quite well, and contemplated marriage.

No. 23.

Chronic and obscure disease of the bladder which had resisted allopathic treatment from the faculty of Leeds, and homœopathic treatment.

J. H. (Leeds), aged 42, married.

In October, 1874, unfavourable symptoms appeared, which, increasing, became very severe. His family doctor said he was suffering from irritable bladder. Medicine doing him very little good, he became impatient, and consulted another doctor, who attributed his disease "as much to indigestion as to his bladder," and promised him "speedy relief." At this time he suffered acute pain—could not urinate at all for considerable intervals, the desire being, nevertheless, very pressing ; had also pain in the lower part of the abdomen ; treated for four months without relief; now sounded for stone but none found. In one of his letters to the writer the patient says :—" The doctor was as much disappointed as I was, and glad, I believe, to get rid of me, as I constantly had to tell him I was no better."

J. H. now consulted a herbalist of known reputation, who told him that he had had "great experience in bladder diseases, and had as many as thirty patients suffering from the disease then under his treatment." This raised the patient's hopes. He took "herb medicine for six weeks," but without the slightest relief. Two other medical practitioners at Leeds were now resorted to. One said his disease was "chronic inflammation of the bladder, which was very difficult and troublesome to cure," and treated the case for several weeks, but without benefitting the patient. The other told him his disease was "gravel," adding "I will soon cure you." J. H. says :—" I carried out his injunctions most faithfully and took all the medicine he gave me without the slightest relief."

Gradually getting worse, and more and more disheartened, he consulted a homœopathic physician in Leeds, who carefully noted all his symptoms and told him, "Your disease has been coming on so long that I do not think I can remove it, but I will do my best." The patient was six weeks under homœopathic treatment without any relief.

J. H., speaking of the homœopathic treatment, says :—" I was in hopes that the *new* system was an improvement on the old, but I was grievously disappointed. At this period I was much worse, which, to me, was a serious matter. I could do little or no work and had a wife and family to keep." Seeing about this time, in the *Leeds Mercury*, an advertisement of one of the writer's books on " Diseases of the bladder cured by a new discovery," he obtained the book, and found cases reported in it resembling his own, and thereupon decided to try what the " spray " treatment would do for him, feeling, as he said, that " it had, at all events, common sense to recommend it, the medicines being conveyed *direct* to the diseased part." He came to London and consulted the writer in June, 1875, the " spray " treatment was applied (though not, of course, exclusively as a specific, as the patient's expression might imply), and in nine weeks he was perfectly cured.

J. H. has continued well up to 1890. In a letter dated March 8th, 1890, he writes :—" It is over sixteen years since I returned to Leeds cured of my painful malady, that had resisted the treatment of eminent allopathic and homœopathic doctors. Since then I have passed through great trouble. Nevertheless, my health has kept well, excepting an attack of colic, and inflammation in my throat. I have had no return of my bladder disease since I left your care. I know, too, of your superior skill in other cases also. Two of these patients whom you cured of stone in the bladder *without cutting* are now spreading your fame and deservedly so. Should any one require confirmation of my statement I shall be pleased to give it for the sake of suffering humanity, and also to testify to you that I am very grateful to you for the marvellous cure wrought in my own case.—I am, my dear doctor, yours very sincerely, J. H."

No. 24.

Obstinate organic stricture of the urethra.

J. A., aged 38, married.

The particulars furnished by this patient show the following :—
After succession of gonorrhœas in early life, he began to suffer
from difficulty in urinating, which gradually increased till the
stream became very small and ultimately dribbled away from him
night and day. He was treated simply as for difficulty in passing
water, without any examination of the urethra being made.
Becoming eventually so ill as not to be able to attend to his daily
duties, he resorted to Guy's Hospital, and was admitted 20th
August, 1879.

Before admission five surgeons had tried to introduce very small
bougies but had not succeeded. He remained in the Hospital until
20th November, 1879 (thirteen weeks), during which time
the surgeons succeeded in getting the very smallest bougie
only into the urethra. Even the smallest bougie introduced
could not be introduced a second time. Numerous surgeons were
taken to see J. A., his case being so unyielding. Indeed one of
the surgeons used to call him the "Curiosity Shop." One
surgeon took him in hand for a fortnight, then another, but
without success. Ultimately one (who had previously treated
the case) succeeded in finding a passage. The bougie was tied
in for four hours, but it "worked out." After another attempt,
for a whole week a small instrument was again introduced, and
kept in for seven days, when this was taken out and an attempt
made to insert another. After unsuccessful efforts made during
another fortnight, other and more violent means were resorted to.
He said :—"Olive oil and carbolic acid were injected and larger
instruments employed by another surgeon. This brought on severe
rigors, and made me so ill that I could hardly stand. I lost my
appetite, and lost flesh ; my temperature went up to 101·8, and
my pulse rose to 130. The case was so severe and unusual that

J

Mr. S. gave a lecture on it." J. A. stated further as follows :—
" At this juncture I felt that they were experimenting upon me.
I was getting worse instead of better, and being just able to crawl
out of bed I came out, and as soon as I could found my way to
Dr. DAVID JONES' Home Hospital in Dean Street, Soho."

J. A. entered the Hospital on December 2nd, 1879. His
symptoms at that time were in perfect accord with the foregoing
statement. After a succession of warm baths, and preparatory
treatment, attempts were made to get through the stricture, which
was at a distance of a little more than five-and-a-half inches from
the meatus. The writer was just as unsuccessful with the usual
method of employing bougies as were the other surgeons, but
by adopting another plan, which he has found successful where the
usual orthodox method has failed, he succeeded in dilating the
stricture sufficiently to admit the urethrotome. The patient was
placed under ether and operated upon, and ultimately, perfectly
cured.

The treatment after operation was very tedious, a succession of
drawbacks having to be encountered. He had cystitis and orchitis
—which occasionally follow the operation of "urethrotomy"—
particularly in bad constitutions such as his. Each attack subsided,
however, under gentle and soothing treatment ; but to effect the
cure nearly three months treatment was requisite instead of the
usual fourteen days, or at most twenty-one days. In the latter
part of 1881 the patient called on the writer and reported himself
as being perfectly well.

No. 25.
Paralysis of the bladder.
S. H. T., aged 35, single.

In this case the patient consulted the writer on the 5th August,
1878. He stated that his general health had been pretty good
from childhood,

On Whit Monday, 1876, he excursioned by train to Liverpool. On the return journey he felt an urgent desire to urinate. No opportunity, however, presented itself, and the discomfort painfully increased. On the train reaching Wrexham he endeavoured to obtain relief, but while urinating he heard the slamming of doors and the railway whistle, and to regain the train hurried back. The train continued its course for six hours, and no opportunity presented itself until he reached his destination, when the bladder would not respond. He went to bed in great trouble of mind, and with considerable physical suffering in the region of the bladder. Next morning he consulted his club doctor, and took medicine for some time, with but little relief. He further described his suffering thus :— " Constant desire to urinate—could only pass a few drops at short intervals—constant burning pain increasing in severity—after a time desire still more urgent, the period between the desire and the act becoming longer."

He now sought further medical aid, resorting from time to time to no less than ten physicians and surgeons, the most notable in Birmingham. Getting worse however rather than better he applied to the Birmingham Free Hospital, but with no better result—being told there that his case was incurable.

Allopathic treatment proving ineffectual, he became an inmate of the Birmingham Homœopathic Hospital and was there two months, and though every attention was paid to his case scarcely any relief was afforded. Medicine failing, galvanism was applied night and morning, but with no decided benefit, and he was accordingly discharged as suffering from " paralysis of the spinal cord and bladder," and " incurable." Again however resorting to the Birmingham Free Hospital, the surgeon told him that medicine was useless, and gave him a catheter to use, without which he only urinated " in dribbles." An operation was likewise performed on the urinary meatus, which relieved him partially for eight or nine days, but he soon relapsed into a "helpless, miserable condition."

When the writer first saw him (August 5th, 1878) he complained of pain in his back ; had an unsteady gait ; said his legs did not respond to his will, and that he had to be very careful in turning suddenly, lest he should fall ; his urine "dribbled from him occasionally ;" had difficulty in starting his urine, and he added " I have a constant desire upon me, but I have no power. I have often had to try for three-quarters of an hour at a time before I could pass any water, and when it comes it does so only in drops." The writer drew by catheter twenty ounces of urine, which was cloudy, very ammoniacal and offensive; administered a " spray" and requested him to call every other day, and to continue the use of the catheter.

August 7th.—He said : " I am better, sir ; I am a great deal easier. I have not such a miserable and constant desire to pass water, and when I have a desire I pass it naturally which is a great comfort to me. I have on that account not used the catheter." Drew by catheter two ounces and five drachms—a great contrast compared with twenty ounces only two days previously.

August 11th.—He stated, " I have passed water naturally several times a day since I saw you."

August 14th.—" I am better, I retain my water longer. I have lost the false attempt to pass water to a great extent."

August 17th.—Urinated naturally in the presence of the writer to the amount of eleven ounces and a half ; drew by catheter only half an ounce.

The reader will note that up to this time the patient had made considerable progress, which so satisfied him that he left London nearly cured. The writer saw nothing of the patient afterwards for nearly five months.

On the 6th of January, 1879, he again presented himself. He said that he had been much better since his last visit in the middle of August, 1878,—" better in every respect." He could pass water with perfect ease, and had not used the catheter since he left London in August, 1878. "Your treatment," he added, " is the only thing that did me any good, but I fluctuate a little, and I have come this time to let you finish curing me. My greatest trouble is the back and limbs. I have to be careful in walking— I stagger, and without care I should be down. I am weak-like in myself, and feel tremulous on the least exertion. I have no spring in me, but as far as regards my bladder, it is wonderful ; still, it takes me a few minutes to start it, but it comes, and in a good stream. Before I came to you it was just a dribble ; in fact, it was endless unless I used the catheter, which I always did." The writer drew by catheter two ounces of urine which had an acid reaction ; it was clear but of a somewhat pale colour—heat made it cloudy. The addition of nitric acid cleared it, showing the presence of phosphates. Gave him a " spray " and told him to call three times a week.

January 9th, 1879.—He steps into the writer's consulting room with a happy countenance, and says : " Your treatment is continuing to act. Yesterday was the best day I have had since I have been ill. I went for three hours without urinating, and when I did so it was quite natural. I only went twice last night." When told to urinate he said, "Ah, that is where I am fixed, you must wait a bit ; it is not long since I passed water." He urinated, however one ounce, and on introducing the catheter about two drachms were drawn.

January 11th.—He came again and said : " I am decidedly better, your treatment is doing its work well. I experience a degree of pleasure in life that I have not done since I have been ill. I passed water only twice last night, and did it like a man, in a natural quantity. When I first came to you I was always at it every hour if you remember, and in perfect misery—never satisfied. The desire was always on me, without any satisfation ; it used to cling to me and never leave me. I have made more progress *this* time than last."

January 15th.—He quite emptied his bladder—not a drop left. Has not used the catheter.

January 20th.—Has continued to improve. Emptied his bladder naturally and thoroughly without any discomfort.

January 29th.—He wishes to know whether he may go home, saying, " I feel quite well, doctor. I have no pain or uneasiness of any kind. I pass water like other people three and four times a day, and once at night, which to me doctor is a great consolation."

February 8th.—Came for the last time. He is quite cured—empties his bladder, being able to urinate freely and fully.

The patient went home perfectly satisfied, and has so continued, as the following correspondence will show :—

"Hawthorne Street, West Smethwick, Staffordshire,
" 10th March, 1879.

" DEAR SIR,—After being at home three weeks, I feel very pleased to inform you that I am getting on very nicely ; and I believe that before long I shall be able to get work. I am feeling some pleasure in life again, for I can venture anywhere now without fear of pain or inconvenience. In fact, I have no pain at all in the region of the bladder. All that I can complain of now is feeling a weakness in the back. I have felt considerable pain at times just below the loins, and I have scarcely got my old strength in my legs, but still I can do wonders in the way of locomotion to what I could previous to being under your care. A great number of people have expressed astonishment to see me go about with the old speed and spring. I cannot express the gratitude I feel to you, sir, for the wonderful difference you have made in me. I can only thank you, but shall always consider you my benefactor and the saviour of my life. With fervent thanks and sincere regards,

" I am, sir, yours truly,

" S. H. T——.

" To DR. JONES."

On the 9th June, 1879, he writes :—"You have my consent to publish my case in any form you think proper. I am quite well, and a living witness to the success of your treatment, when neither allopathic or homœopathic treatment was of any avail."

In a further letter dated 14th December, 1879, the patient verifies the accuracy of the foregoing statement of his case (a copy of which had been sent him for perusal), and mentions that he is in suitable employment, and has "felt better in every way since he commenced," experiencing occasionally a "slight weakness" but "no pain or inconvenience from the bladder."

The subjoined letter authenticates the above. The copy of it was handed to the writer by the gentleman who had received the letter from his son-in-law, as the result of enquiries made by him in reference to the case.

26th April, 1879.

"MY DEAR FATHER,—I have just returned from seeing T——. The Vicar of —— had not returned from Leamington. It was not, however, of importance, as T—— has long been connected with the Rev. E. A——'s congregation as a member of the choir. It was Mr. A—— who helped to send him to London. T—— showed me a list of subscribers to the fund to send him to London. He is a very intelligent man, aged thirty-seven. His disease arose from too long retention of urine when on a rail journey. His symptoms were very like yours. The letter to Dr. Jones is genuine. He is quite in his old usual state as respects his bladder, and says he is gradually gaining strength. He looks very well. He is told that what he is now suffering from, and that slightly, is connected with the spine. His description of his painful feelings, in trying to make water, is much like what you have told me of yourself, only he says he never discovered any appearance of blood; but that may be accounted for by the fact that he used to sit in the water-closet for an hour at a time, the water just coming in drops.

He described the sensation of ease after the first operation as something delightful. He says the operation did not give him much pain. He strongly advises your going immediately to Dr. JONES, and I would do the same. By delays you are only risking the desired-for success of the treatment, and continuing your pain."

S. H. T. continues well up to the present time. May 1890.

No. 26.

Disease of the bladder, with enlarged prostate—catheter wholly discontinued after having been continuously used for upwards of two years.

S. M., aged 66, gardener.

This patient consulted the writer on the 7th November, 1877. He had been ill for more than two years, his urinary discomfort having commenced with more frequent desire to urinate than usual, ultimately however so increasing as seriously to affect his occupation during the day, as well as his rest at night. He said:— "My urine became hot—it cut me like a knife—it got thick like glue. The doctor relieved me with instruments, for a time, but my disease getting the upper hand he sent me to Guy's Hospital, where I attended every fortnight for two years, and was there taught to use the catheter, which I have done regularly ever since."

An examination revealed an enlarged prostate—urine highly alkaline and loaded with mucus—the aperture of the urethra very inflamed and swollen. The writer drew by catheter twelve ounces of thick ammoniacal fluid, administered spray treatment two or three times a week, and advised disuse of the catheter. Each time the patient came he reported himself better.

On the 5th December (just a month from the time when first seen) he stated:—"I think I am well sir. How pleased I am that I heard of you. Just think of the length of time I have been to and fro to the Hospital. I now only pass water three or

four times a day, have no pain, and my water is quite clear."
Residuary urine found to be not quite two ounces, and normally
acid. The spray treatment again applied at intervals for a
fortnight, then discontinued.

On the 12th of June, 1878, S. M. called on the writer and said :
" I am quite well, have no urinary trouble of any kind, and am
quite free from pain." Drew however an ounce of residuary
urine, normally acid. No further treatment recommended.

On the 7th June, 1882, the writer again saw S.M., who then
said : " I have been perfectly well since you cured me in 1877,
but my poor master is dead. It is curious sir that (as far as I
could learn) he died of a similar complaint to that I suffered from
myself. I had told the butler that master knew how bad I was
and that Dr. JONES had cured me. Some London physicians
were however sent for, but unfortunately did him no good." He
added :—" I met my old doctor three months back ; he asked me
if I still used the catheter. I said No, and that I should have
died had it not been for Dr. JONES."

On being questioned, S. M. assured the writer that he was as
well and as free from pain as he had ever been in his life, and
that he had not suffered any urinary discomfort since he left off
treatment—more than four-and-a-half years ago—nor had he
again used any catheter. On examination *per anum* the prostate
was found still large, but there was no tenderness on pressure.
On being asked if he had any rectum discomfort, he replied that
the uneasiness there was cured when the bladder was. This
patient expressed his willingness to have his case published and to
answer any questions any enquirer might wish to put in reference
to it. He has been gardener, in the employ of the same family,
for more than thirty-five years, his present employer being a
clergyman, son of his late master, and who now resides in the
same house.

No. 27.

Paralysis of the bladder.

H. D., aged 45, married.

This case illustrates *paralysis* of the bladder, greatly relieved—apparently cured—various previously applied means having failed to take effect, galvanism included.

This patient consulted the writer on the 13th December, 1877. Nine months previously he had been seized with loss of power in his lower limbs and had been treated in a hospital and at other places for paralysis. Soon after was seized with considerable pain in his back which increased the paralytic sufferings, and occasioned inability to retain his urine, which (as he said) "dribbled away night and day." The like loss of power extended to the bowels, and greatly added to the patient's discomfort.

When first presenting himself he was in a most deplorable state and having no urinal, flannel and similar contrivances were made use of. Every one he approached perceived an offensive odour. He had extreme irritation of the bladder, and was obliged to urinate in very small quantities continually, and without much relief. On introducing a catheter, thirty ounces of highly offensive turbid urine loaded with mucus were withdrawn, and a "spray" was administered. He was instructed to wear an urinal (of which he had not previously heard)—was treated twice a week, and made very rapid progress.

December 24th.—Residuary urine, heretofore very muddy thick and offensive, now clearer and inoffensive, and reduced from thirty ounces to four ounces. Constant urination not so frequent, and feels in every way a better man than he had since his seizure. For safety administered another spray.

January 3rd, 1878.—He declares he is cured, urinates naturally, and says: "I can do without the urinal night and day." No treatment was used, and was told to come again in a few weeks.

January 21st.—Came to the writer for the last time, considers himself cured as far as the bladder is concerned. Discontinued treatment, but advised to come again if necessary.

May 25th.—Came to the writer and reported himself " quite well in the bladder," but added—" My limbs are still very weak. Can't you cure my legs now that you have made such a wonderful job of my bladder ? " No treatment on this occasion.

October 23rd.—Came again—slight return of bladder irritation —says :—" I have done my work very well since you cured me, but I am afraid of this scalding." In answer to a question respecting the urinal, he said :—" I have had no occasion to use it since I left it off." Urinated naturally six ounces of clear urine.

November 6th.—He says :—" I am very much better sir. I think I shall do now." Emptied his bladder naturally. On introducing the catheter no residuary urine was withdrawn. No spray used—was told to come again in twelve months.

October 16th, 1879.—Called and said :—" I have followed my occupation without interruption as far as my bladder is concerned." H. D. gives the same report up to this date, 1890.

NOTE.

The above case is not one presented as perfectly cured, but as illustrating the ameliorative effects of the writer's treatment. He is not aware of any case of the kind in which any real or lasting benefit has been derived from the ordinary method of treatment, all such cases being invariably regarded as beyond relief. Under the ameliorative treatment alluded to, the paralysis and inflamma-tion of the bladder consequent, upon the retained and decomposed urine, are markedly subdued, as shown in the foregoing case— the helpless, paralysed condition of the bladder soon responding to the treatment. A thorough and permanent cure cannot, however, in all cases be expected, for the brain and spinal degeneration continue to get worse ; still, the patient's life may be prolonged, his sufferings greatly mitigated, and comfort and consolation derived from the conscious ability to pursue the active duties of life. In the above case the patient was enabled to earn the means of subsistence for himself and his family. He had, moreover, the satisfaction of seeing his children grow up until capable of helping themselves, which was in itself an inexpressible joy to him.

NOTE.

The subjoined cases illustrate the cure of disease in the *female* bladder, etc., and the attention of the reader is desired to the following preliminary observations—applicable more especially to cases where the disease is connected with sterility.

These cases are often very complicated—having a variety of causes.

1. Irritability in *unmarried* women, when *unattended* with pain, but attended with frequent micturition, (the urine being as a rule clear and not indicating inflammatory action) is generally sympathetic in its origin. Such cases are usually complicated with hysteria, and the primary cause is essentially uterine.

2. Irritability of the bladder in *unmarried* women *attended* with constant and painful micturition *without* the usual manifestations of hysteria, *i.e.*, urinary secretion clear and watery, without inflammatory indications. In these cases there is a morbid sensitiveness of the spinal cord, at the spot where the uterine nerves are given off. The symptoms are due to uterine disturbance, hence the treatment must be directed to the general health, spinal, and uterine systems.

3. Irritability of the female bladder in *married* women who have *not* borne children, or who are married to old men, and are also sterile.

4. Irritability attended with considerable pain of an inflammatory kind, where the urinary secretion is cloudy and sometimes bloody. These symptoms appear in sterile women, married to

men of suitable age, and are due solely to uterine irritation—a condition which the writer has called " disappointed womb." It is only curable by removing the sterile condition.

The writer has frequently met with cases of the kind above referred to, which he has cured by removing the cause, without treating the bladder at all. The bladder symptoms are as a rule purely sympathetic.

The writer has met with such cases where all the usual forms of medical and surgical treatment had been applied—and in no instance has he been more *unsuccessful* than in cases where the treatment called " rapid and forcible dilatation" had been adopted. He recalls four cases—two in Leeds, one in Newbury, and the other (wife of a physician) in Birmingham. In two of them, the patients had been literally lacerated from the entrance of the urethra to the neck of the bladder, and rendered miserable for life. In both the urine dribbled away night and day without the slightest power of control.

In addition to the foregoing, there are cases of disease of the female bladder of a painful and inflammatory kind, occurring during pregnancy.

These cases are, as a rule, seldom cured by the modes of treatment ordinarily adopted. The consolation which patients get is, " Well, Mrs. ——, you see the irritability is due to your pregnancy, and will pass off when you are confined." They are treated by palliatives—opium, morphia, belladonna, etc.,—and by suppositories, warm baths, etc. Unless relieved by miscarriage, or premature labor (a common occurrence) the patient drags on a miserable existence until her confinement is over. No doubt in many cases patients get better after this, but where such treatment has been applied, there is always liability to a return of the disease, especially on exposure to cold, and cases of this kind not infrequently become chronic. So severe a form of the disease is however not of very frequent occurrence. In cases treated by the writer, one was under treatment for six weeks, and was cured within that time — another in five weeks — a third in four weeks and two days—a fourth in three weeks, and the last case (now very advanced in pregnancy) was discharged cured after four weeks' treatment, but the writer feels, at present, a little uncertain, whether this favourable condition will continue until the end of gestation. In the first four cases, the patients continued well till after they were confined. One of them has since had three confinements. Another three, and another one, and all without any return of their former distressing symptoms.

Most pregnant women suffer more or less from irritability of the bladder during the early months, the discomfort being merely a frequent desire to micturate, without pain however of an inflammatory character. Such discomfort is bearable, and is to a great extent, under the patients' own control, and requires no particular treatment, save abstinence from fluids and stimulants. As pregnancy advances the gravid uterus is elevated (so to speak) from the pelvis and ceases to press on the bladder, and the irritation ceases.

But the cases now to be alluded to differ very materially from those first referred to in these observations. In the following it was not only desirable for the comfort, but important to the health of the patient, that relief should be afforded by the application of a treatment really curative.

No. 28.

Mrs. R., aged 29.

The following is a case illustrating uterine irritation in the *unmarried*, as well as in the *married sterile* female.

The discomfort commenced before marriage, establishing one set of symptoms, and continued after marriage, establishing a different set of symptoms, both nevertheless due to uterine sympathy. For many years she suffered from frequent micturition unattended by pain, the urine being clear and *not* inflammatory, as in unmarried women predisposed to nervousness and hysteria.

Three years previously to 1879, when the writer was first consulted the irritation suddenly became worse, culminating in a slight discharge of blood in the urine. This later symptom soon yielded to mild treatment, the patient being left with merely frequent micturition without pain.

She married in 1874. Eight months afterwards she suffered pain of a more decided character extending from the region of the bladder to the left hip. Soon after she took a violent cold which aggravated her bladder discomfort, and brought on bearing down and forcing pains in urinating. She had cold shivers, alternating with heat and fever. Dr. D., a neighbouring prac-titioner was sent for, but treatment was of no avail. Mr. C. H., hospital surgeon, now held a consultation with Dr. D. and said "she had a florid growth in the urethra," and removed it under the influence of ether which did her good for a time—former dis-comfort soon however returned, and her old symptoms of forcing, burning and intolerable pain were as violent as ever. "All these doctors did me no good," was her complaint. Another physician's treatment also proved fruitless of benefit.

When the writer first saw her, she had been confined to her bed for five months. She was considerably emaciated—had a small and almost imperceptible pulse, 110 in a minute—urine turbid and albuminous — micturition constant and painful, and passed with spasmodic violence. She said, " the urine stops every now and then, and comes on again." The writer diagnosed the case as due to a "disappointed womb," and treated it accordingly.

After about a fortnight's treatment she was able to get out of bed a little, and in three weeks she was removed to the Home Hospital, 10, Dean Street, Soho. Under the treatment here applied her severe symptoms soon subsided—she visibly improved in general health—her local symptoms disappeared one by one— spasmodic forcing pains first gave way—nights improved, slept for long intervals at a time—appetite returned—lost the yellow cast of countenance borne for so long a time, and in fact health was restored. She soon left the Home Hospital for her own home in Switzerland, where she has been ever since in the full enjoyment of health.

In less than twelve months after her return, the writer received intelligence that she was safely delivered of a healthy child. Since then she has had a second child, and just previously to April, 1882, the writer heard from a friend of hers (Mrs. C——, Peckham), informing him that another addition to her family was then expected.

The reader will see from this case that in the first instance the old routine treatment was adopted from beginning to end— medicines for the bladder (which was only sympathetically affected), removal of a " florid growth " which was the effect and not the cause, and sounding for stone by a surgeon of one of the principal of our metropolitan hospitals on the principles taught 30 years ago.

The following, extracted from the patient's letter, bearing date 13th March, 1881, corroborates the writer's successful treatment.

" Switzerland.

" I left your Home Hospital in 10, Dean Street, on the 13th September, 1879, feeling confidence in what you told me, which came quite true. We left for Switzerland soon afterwards. At the end of October I found I was *enceinte*, when my bladder became a little troublesome, which no doubt was caused by my condition. I was confined on the 29th of June, and my bladder became better. Unfortunately I was torn in my confinement, which has made me think that my bladder might be a little troublesome through that circumstance. November, 1880, arrived, and I was again *enceinte*. I often think of your telling me that "all nervousness would disappear if I had children," and so it has, but my kidneys are bad, which makes me feel I am not *quite* well. I must tell you that you *alone* had the talent to replace me on my feet after five months passed in bed. Hoping that this letter will find you in good health, I beg you dear Doctor JONES, to accept the assurance of my high regard, and to receive the best wishes of my husband and self. "Yours sincerely,

"J—— R——."

—————

No. 29.

Uterine misplacement (mistaken for tumour) producing disease of the bladder, and sterility.

E. R., aged 28, married.—Sterile for three years.

This patient dated her discomfort from an early age. When a young girl had been assistant to a greengrocer and had to stand about for many hours every day, and lift heavy weights. The labour connected with this led to serious disturbance in her monthly courses, associated with constant irritability of the bladder and repeated attacks of sickness. The bladder irritation though very inconvenient, was not accompanied with much pain until she was married (at the age of 24) when serious symptoms quickly developed themselves, married life being extremely painful. Medical advice and assistance were sought. The ordinary family doctor, and subsequently several others were consulted, but without relief. Then the Hospital for Women, Soho Square, was resorted to. Her case was there considered a serious one. The medical men attached to that institution all agreed that she was suffering from ulceration of the womb—one of them also affirming the existence of a tumour "as large as an orange." After remedies applied for two months, she was advised to enter the hospital as indoor patient, to which however she

K

objected. Next Bartholomew's Hospital was resorted to, the physician there (a specialist in diseases of women) saying it was lucky she had not impregnated, adding that she had a tumour and that there was no room for a child to come away. She received however no benefit from his treatment—his successor at the same hospital being equally unsuccessful, and he also said that she had a tumour as large as an egg. The writer was consulted on the 14th August, 1878, three years after her marriage. Her symptoms at the time were (1) bladder irritation gradually getting worse, signs of inflammatory pain, frequent urination with scalding pain, urine cloudy; (2) constant pain in the region of the bladder; (3) a gnawing pain extending to the hips and loins; (4) constant nausea and dizziness of the eyes; (5) constipation attended with pain when the bowels were opened.

A careful internal examination revealed a swelling the size of a small orange in the hollow of the sacrum, while the neck of the womb was turned upwards and forwards. It was this swelling which had been mistaken for a tumour—it was in fact the fundus of the womb pressing on the bowel, while the neck of the womb pressed on the bladder, occasioning the discomfort complained of. The treatment adopted by the writer was directed to the removal of the sterile condition of the patient, the womb was replaced by gentle means that is to say by gradually levering it to its proper position by means of cotton pads (not by pessaries as usually employed in such cases) and by attention to position. In the course of three months the unfavourable symptoms disappeared. In the fourth month signs of pregnancy presented themselves. This also contributed to the lifting of the womb from the pelvis, so that the bladder and lower bowel were left undisturbed,—in other words the patient was cured. In due course she was safely delivered of a healthy child and made a good recovery.

Since then E. R. gave birth to a second child, and is now (1882) expecting a third.

The above lady has lived in the same house in Bedford Row, Holborn, for twenty years and will gladly verify the above statement if requested.

No. 30.

Chronic and " incurable" disease of the bladder, previously unsuccessfully treated in Russia, Germany, France and England, by allopathic and homœopathic physicians.

S. M., a lady advanced in years, was travelling from Paris to St. Petersburg by train in 1863, when a severe frost set in for which she was not suitably prepared.   The sudden reduction of temperature during the journey occasioned—contrary to her usual habit—a painful desire to micturate (empty the bladder). To effectuate this during one stage of the journey was however impossible, by reason of the distance to the next station.   On arriving at St. Petersburg considerable uneasiness and painful irritation was experienced, but ultimately after considerable difficulty relief was obtained.   The consequence of this retention was the establishment of constant irritation at the neck of the bladder—forcing pain and false attempts to urinate.   This discomfort continued, despite the aid of able professional men. Various kinds of medicines prescribed by physicians gave only temporary relief.   A change to Germany was now recommended, and a physician of the University of Heidelberg was consulted. Stone in the bladder was supposed to be the cause of the continued irritation but an examination by " sounding " proved that such was not the case.   The professor suggested various remedies and soothing fomentations.   These failed even to give relief.   The waters of Ems were suggested, and the most able physicians there were consulted.   A long course of treatment resulted in similar disappointment.   The next step was to France —Vichy waters were strongly recommended—a noted physician

was consulted, and his advice strictly followed, but with no better success. Next a physician of world-wide reputation, in Luchon, in the Pyrenees, was resorted to, who prescribed sulphur baths and fomentations of famous herbs. Notwithstanding all these efforts her malady still gained ground and for some time she was unable even to walk without pain. Coming to England in 1865, she consulted several physicians of eminence, was examined at repeated intervals, and treated for "stone"; but still without any success. Being strongly advised by one of these English physicians to go into St. George's Hospital, she did so; and having been admitted there, underwent several painful examinations, and had the advantage of consulting the medical and surgical staff. Of her hospital life the patient thus writes— (and it may be here mentioned that, following the plan adopted in compiling these cases, the foregoing as also most of the subsequently stated particulars are taken from her own narrative) :—" I conformed to the hospital rules for some time, but was no better; and the physicians failing to do me any good, were glad, I suppose, to get rid of me, and dismissed me, as my former advisers had done." She then proceeds to narrate the circumstances under which, with forlorn hope, and in the midst of much solitary suffering she consulted another physician whose advice she followed for about nine month, but without success. Then homœopathy was recommended, and with hope raised she consulted eminent physicians of that school,—still however without beneficial result. After this a medical practitioner at Uckfield in Sussex was consulted who liberally administered morphia, iron and blue pill. She found herself nevertheless getting worse instead of better. Her narrative then proceeds :—
" My suffering changed its character. Every now and then I was seized with spasmodic or cramp-like paroxysms of intense agony which made me scream. This would continue from one, two, and even three hours. At this time I hardly ever slept half an hour together,—the extreme urging to pass water awoke me from 15 to

20 and more times during the night—and not unfrequently during the day I was troubled in the same manner, 10 or 12 times in an hour." Then, after detailing the hopeless and almost despairing condition of mind into which she had fallen, and the circumstances under which she had been led to consult the writer, she says : " I paid Dr. D. JONES a visit, who carefully noted all my varied symptoms ; and the better to secure his personal attention I entered his Home. The first week passed without apparent difference as to suffering but after that time I improved so rapidly that in three weeks and three days I went to church for the first time for years. The pain and all its inconveniences left me ; no vestige of past suffering remaining. Thus contrary to hope (humanly speaking) Dr. JONES was the instrument in God's hands of curing me after eight and a half years continued suffering, and after so many attempts of other physicians, of unquestionable repute in Russia, Germany, France and England."

This lady continues perfectly well now (1882), more than eight years since she left Bolton House, and has had no return of her disease. She will willingly verify the foregoing statement, and if requested will furnish the names of the several physicians whom she had consulted. Since then she has undergone an operation for cataract by an eminent oculist.

No. 31.

Chronic and supposed incurable disease of the bladder, cured in six weeks.

Mrs. E. M. A., aged 54.

This case was very nearly identical with the case of S. M., above described. The symptoms were as severe, the duration of suffering was 18 months more, its incurable character was equally marked, while the exciting cause was entirely different. The patient resided at 164, Piccadilly, premises known as "The Religious Tract Society."* She had lived there for upwards of 20 years, and is well known to the Secretary of the Society and the neighbours. Her suffering commenced 10 years ago, during the "change of life," she being at that period 44 years old. The exciting cause she attributes to tripping accidentally against a door mat, and falling down stairs, which shook her considerably and brought on a pain in her side, as well as discomfort in the lower part of the abdomen. This was soon followed by irritation in the bladder at irregular intervals. Soon after, the irritation became more troublesome. Mrs. A—— describes her sufferings thus : " I had to pass water at first three or four times an hour, which was attended with acute forcing or bearing-down pain, as if my inside would be forced out." " These inconveniences," she adds, " became more and more troublesome, the bladder having to be emptied every few minutes. For several years, at intervals, the irritation and constant urging to pass water were so severe that at night no sooner was I in bed than I had to get up again. Finding it useless to try to get a little sleep, I adopted the plan of propping myself outside, and on the edge of the bed, wrapped in blankets, with the chamber utensil conveniently placed ; by this arrangement I was not so thoroughly roused from sleep. When going out of doors, I had to make preparations for it, and was compelled during the paroxysm of pain to cling to posts or railings, or anything I could get hold of. Occasionally I had to sit down on the doorsteps, screaming with pain." In another part of her narrative she states: " My poor husband was so anxious about me that he spent over 100 guineas in obtaining the advice of the most eminent physicians in London, who variously described my disease as inflammation of the neck of the bladder—chronic disease of the mucous membrane of the bladder—stone in the bladder—impacted stone—tumour pressing on the bladder—falling of the womb—gravel—disease of the nerves of the bladder

---

* For Mrs. A.'s present address see the Appendix.

—ulceration of the lining membrane of the bladder—malignant disease of the bladder, &c., &c. All agreed that I was incurable." This patient (from whose narrative as in the former case these particulars are taken), afterwards became a patient of the writer. She was under treatment for six weeks, was discharged cured, and continues well without the slightest return of her old symptoms. She writes in a book recently presented to the writer, as follows :—" The gift of a grateful patient, E. M. A——, who had suffered from an internal complaint for 10 years, had been an in and out-door patient of five hospitals and unsuccessfully treated by upwards of 20 physicians, and was cured by Dr. JONES in six weeks."

The hospitals to which Mrs. A—— resorted were, Guy's, St. George's, Middlesex, Hospital for Diseases of Women, Soho Square, and the Samaritan Free Hospital.

NOTE.

In this case, the writer did not in the first instance apply his own new method of treatment only, but the specifics (so called) in the allopathic and homœopathic pharmacopœa, though without success. As a homœopath, he can affirm that full justice was done to homœopathy and homœopathic remedies as used in such cases, and in further justice to allopathy, it may be stated that the old system was (as appears from Mrs. A——'s own statement) fairly tried by some of the most eminent practitioners in England; concerning whom, and her case generally, Mrs. A—— has expressed her readiness to furnish any information which may be desired.

No. 32.

An obscure case of bladder disease resembling cancer, previously unsuccessfully treated both by allopathic and homœopathic means.

MRS. S. C., aged 53, married.

October 14th, 1875.—Her own statement of her case was substantially as follows.—She had been suffering for five years and a half, her disease being, as she believed, exactly the same as that described in Mrs. E. M. A.'s case, which case had come under her notice in perusing a previous edition of this work. The doctors differed in opinion as to the nature of her disease, and told her that she could not be cured. She was treated by an experienced homœopathic physician for twelve months. Getting no relief, she went to the Samaritan Free Hospital for several months, without benefit. She afterwards went to St. Peter's Hospital for twelve months, still she got no better. Lastly, she went to another hospital (one inaugurated by a lady M D.) for four months : one month as in-door, and three months as out-door patient,—the result being negative, but rather an aggravation of her sufferings. Since the commencement of her disease she had been gradually getting worse. She had felt wretched and miserable in herself and was a trouble to others.

Mrs. C.'s appearance presented the characteristics usual in such cases. She looked pale, dejected, and worn out with pain, owing to an almost ceaseless straining effort to urinate. The writer's note-book records the following symptoms when first he saw her :—"(1) has emaciated considerably during her illness, countenance sallow, resembling patients dying in the last stage of cancer—(2) pulse very small, weak, and frequent—(3) tongue furred on the dorsum ; the anterior portion, and the edges, extending as far as the root, are red, irritable, and cracked—(4) has constant nausea and disagreeable taste in her mouth—(5) the urethral canal, extending to the bladder, is nearly closed ; and it is with difficulty that a small catheter can pass; it is also thickened considerably (as thick as a thumb), and feels as hard as a cord, and very painful on pressure ; the bladder bulges into the vagina, forming what is called ' cystocele' (projection of a portion of the bladder into the vagina)—(6) has considerable pain over the regions of both kidneys extending along the course of the ureters into the groins—(7) she can only partially empty the bladder—(8) the urinary secretion is cloudy and deposits a thick sediment which under the microscope proves to be pus and mucus with a large quantity of spheroidal and tesselated epithelium with crystals of triple phosphates of ammonia and magnesia, such as is frequently seen in prostatic disease in the male—(9) the urine highly alkaline and albuminous—(10) has most of the symptoms complained of by Mrs. A.," (see Mrs. E. M. A.'s case). The wide-spread implication of the whole viscus had occasioned extension of the mischief into the ureters (the tubes which convey the urinary secretion from the kidneys into the bladder), and from the ureters into the interior of the kidneys. This was evidenced by the albuminous condition of the urine. The kidneys were rendered incapable of removing urinary products from the blood. These irritating secretions were partially retained, occasioning the constitutional symptoms from which the patient suffered ; in other words, there was a degree of uræmic poisoning which is common in " Bright's disease of the kidneys," which proves fatal in thousands of cases annually, and to which Mrs. C. must have succumbed had not the *cause* of her disease been discovered and removed. Clearly the above case presented very grave peculiarities and complications, and was (so far as any ordinary plans of treatment went) utterly hopeless, and must soon have proved fatal. This conclusion was fully justified by the then unfavourable results. Allopathic and homœopathic physicians alike pronounced her incurable, and her bountiful friends beginning to feel the drain on their resources, threatened to withdraw their help when they saw that no benefit was derived from their generous outlays.

Under these circumstances, and the patient believing the case to be within the reach of the writer's treatment,—a belief in which he greatly encouraged her,—he gave her a free admission to his Private Hospital for Women. She entered, much against the wishes of her friends, and to their astonishment, returned home in four weeks cured. After a few weeks she fancied she had a slight relapse. Two more applications with the " spray," however, completely cured her, and she continued well up to 1881.* When once the *cause*, which was in the bladder and urethra was removed, the kidneys, which were only *secondarily* affected, soon got well· The back pains complained of shortly vanished, and the urinary secretion quickly gave evidence that the kidneys were doing their work healthily ; and as a general result, the haggard, worn-out aspect gave place to a healthy hue and happy countenance.

This case is well known to a large number of friends at St. Peter's Park Baptist Chapel, the minister of which is ready and willing to verify the above statement if requested to do so.

No. 33.

Chronic bladder disease of six years' duration, occurring after a confinement.

F. T., aged 28, married, and mother of four children.

April 12th, 1880, stated as follows :—" I have been ill since my confinement six years ago, and have been to several doctors who have done me no good : they do not seem to know what my disease is. It came on three days after my last confinement. My own doctor treated me without success for some time—when, thinking myself a little better and trying to move about I found myself worse. The doctor tried a great many different remedies but without avail. I was now sent to the Hitchin Infirmary, but I got worse the whole time I remained, viz., three months. Deriving no benefit I thought it was no use to continue any‘ longer. I suffered great pain, and had to pass water every few minutes night and day. I lingered on in this way for two years, and being only a poor man's wife I felt it all the more, as I had to neglect my husband and children. Getting worse and worse,

* On two or three occasions she was somewhat inconvenienced by *another* disease.

the rector of our parish took my case up, and was very kind to me, and wrote me a letter to take to the doctor I had been under for three months on a former occasion. He paid me every attention, and said he had done all he could for me and hardly knew what was the matter with me. I was under his treatment on this occasion for two months longer, still without relief, and accordingly left off doctoring again. I lingered on this time in great agony for another year, and became pregnant again. I suffered at this time more than ever. I thought I must have died before my confinement, but everybody kept telling me I should be better when the child came. I got over my confinement better than I expected, but my sufferings were very great day and night.

At the end of three months I got much worse and was advised to go to another doctor in Hitchin. I told him the doctors I had been under, and asked him if he could really tell me what my disease was. He replied that he thought it must be severe ulceration of the bladder, and he examined me very carefully on two occasions during the five months I was under his care, but he did me no good; no one did me any good; I did not even get relief. I think I became worse after the treatment of every doctor I was under. At this time I had more burning pain than ever—I can only compare the pain as if boiling water was passing from my bladder. I had sometimes to go every three or four minutes, but usually six to seven times an hour. After five months doctoring I tried another doctor living in Halsey. This doctor thought I had stone, and examined me for stone twice but found none. I gave him three weeks' trial, and took his medicine regularly without any relief. He said he did not really know what my disease was—the disease was "a very curious one." It is just two months since I left off his treatment and have come to you to see what your skill will do. I have no faith in you or anyone else now, but as The Rev. Mr. M—— urged upon me to come, I have no objection to let you try what you can do."

The writer after making notes of the case in his case-book, administered the first "spray," but before doing so, the patient was asked to empty the bladder, which she tried to do. Catheter introduced—eight ounces of muddy and ammoniacal urine drawn which upon standing deposited 50 per cent. of mucus and blood. The clear fluid contained 5 per cent. of albumen, the usual test of heat and nitric acid having been applied.

April 15th.—Second interview. She said, "I am really better, I slept more last night and the night before than I have since my illness, I am better also in the day time." Drew by catheter same quantity of urine of same character as before, but clearer, and on measurement it yielded only 30 per cent. of mucus or muco-pus.

April 19th.—Third interview. She said, "I am so thankful to tell you that I am getting well. I slept from 10 o'clock last night till 4 this morning, and when I passed water it was clear and gave me no pain in passing. I have also been out for a walk with my sister and walked a mile and a half without much fatigue. I only passed water four times the whole of yesterday, and this I did without any pain." Emptied her bladder—catheter introduced— about an ounce of urine only remained—the urine still a little cloudy—slightly acid—deposited 1 per cent. only of mucus—only slight traces of albumen—scarcely distinguishable by nitric or picric acids.

April 22nd.—The patient states she is "quite cured, I slept from 10 o'clock till half past 7 in the morning, which I have not done for years. I go out every day and walk like anyone else"— told to empty her bladder—this done, showed that the function of urination had quite returned. She said, "the urine is perfectly clear, I sleep all night without being disturbed and I do a great deal of walking during the day." The patient was told to come again in about a week.

April 29th.—The urine examined—normally acid, and of the specific gravity of healthy urine. She walks out daily, eats her food naturally and has gained flesh. Discharged cured (in less than three weeks) but for safety advised to remain with the friend she was staying with for a week longer, and to call on the writer in case of any of her former symptoms returning. She accordingly remained with her friend, Mrs. M——, Victoria Road, Holloway, for about a week.

The patient's friend, whose address is appended above, will be glad to verify the truth of the above statement if requested to do so.

The following is a letter received from the Rev. Mr. M——, the gentleman who sent the patient to the writer :—

"October 20th, 1880.

"My Dear Sir,

"The young woman you so kindly treated at my request wished me to write and inform you that she continues *quite well*, and has not had a *single pain* of the kind for which you treated her, since she came home. Full of gratitude for the wonderful cure so quickly wrought, she is anxious that you should make some use of her case, for the benefit of others. She also wishes me to say that she would be pleased to answer any questions, go anywhere, or do anything in her power to induce any poor sufferer in a similar condition to try your skilful treatment.

"It is certainly a very remarkable case. For six years her sufferings were most distressing, no relief could be obtained either at home, or at the hospital. When I wrote you her case appeared hopeless, and when at the end of one month, you sent her home cured, I was still afraid the relief was only temporary. She is, however, *perfectly* well, and says she never enjoyed better health in her life than she does now. I have known the case from the commencement, and if you are disposed to publish any statement respecting it, shall be pleased for you to make use of my name, or to answer any enquiries that may be addressed to me.

"I am Dear Sir, yours sincerely.

"D —— M —— .'

Some time having elapsed since she returned home, it was thought desirable before going to press to communicate with the gentleman who sent her to the writer, to know her present state of health. The following is a reply.—

"January 19th, 1889.

"My Dear Sir,

"I found F. T. *quite well* and at work. Her case is so complete that she appears years younger than she did, and tells me there has not been the *slightest* inconvenience since she came home. She may well be so desirous of inducing any other poor sufferer to come to you. She is daily expecting to be confined.

"I am, sincerely yours,

"D—— M——."

No. 34.

Stone in the female bladder weighing two ounces and three quarters.

E. H., aged 37, married.

This case illustrates in a marked manner the importance of careful diagnosis—as to which too much is perhaps sometimes expected from the *general practitioner*, who, having all kinds of diseases and ailments to deal with, has little or no time for devoting attention to any one in paticular, as a speciality. But the like allowance can scarcely be made in favour of hospital treatment, where one might reasonably expect that special diseases would be treated by physicians or surgeons specially qualified. Though even there (as happened in the present case) a sufficiently careful and accurate diagnosis is not always observed.

This patient had been ill off and on for five years, her illness culminating in a very severe form of suffering. She was at first attended by a medical gentleman practising in the vicinity of her own residence, who attributed her suffering to "internal abscesses or misplacement of the womb," and advised "instruments," and sent her to Finsbury Pavement to procure the requisite appliances. The instrument used gave her excruciating pain for three weeks; it occasioned moreover the passage of more blood. Getting no relief the instrument was removed. She was now advised to go to St. Bartholomew's Hospital, which she entered as an indoor patient. There her suffering was attributed by the assistant obstetric physician to uterine disease, and treated accordingly. Several pessaries were introduced, but these only intensified her sufferings. Here, too, they drew her urine by catheter. She said to the writer, "I was taken on four occasions to the Theatre of the Hospital on Tuesdays and Fridays in the presence of several doctors." After the *first* examination she got worse, and had to get up twelve times in the night. The nurse was told that her disease was inflammation of the bladder. The second time, when she went to the Hospital Theatre, she was again examined before about twelve doctors. She says :—"Dr. ———, when asked what really was the matter with me, said he thought it was stone, but the stone was too large for operation, and that I would do just as well at home as in the hospital." She had been in hospital a month and three days and got no better. Went again a fortnight after being at home and saw Dr. ——— in the presence of four other doctors. She asked him if he thought it was stone. He said, "Who told you so? it is no stone at all," and added, "take all the rest you can." The other doctor had before told her it *was* stone.

Entering subsequently the Hospital for Women, Soho Square, the physicians there took the same view of her case as had been taken at St. Bartholomew's Hospital, and she was treated for "disease of the womb." The womb was said to press on the bladder. Leeches were applied to the "congested womb" and medicines prescribed. After five weeks treatment under a physician there, and getting worse instead of better she declined to continue any longer under treatment. Her suffering increasing very considerably, and acting under the advice of friends, she again went to St. Bartholomew's Hospital. Dr. —— and several other doctors saw her. She says :—" They seemed cross when I still complained of the water, and said ' when the womb gets well the other will." When I complained in the same way in Soho Hospital they said the same thing. I took my water three times to Soho Hospital and urged upon them to examine the bladder, but they said it was the womb. The instruments made me worse, but all along when I assured the doctors I thought it was the bladder, and not the womb, they said, ' Do you think you know better than we do ?' I said (crying), ' I am sure it *must* be the bladder and not the womb.' The two doctors laughed at me and told me to ' go away ' and that I would ' soon get better.' "

Her suffering now became intolerable and in great despair she left the hospital and went to her home. Becoming impressed with the notion that she was pregnant, the family doctor was sent for, and a very skilful physician's advice was sought, who said that pregnancy was the sole *cause* of her suffering, and gave instructions for premature labour to be brought on forthwith. The physician was evidently under the impression that a gravid womb pressing on the bladder accounted for her suffering, and that bringing on the labour would remove the cause. At all events both patient and friends were assured that it would result in her being cured. During the time this was going on, the bladder was to be relieved by catheter three times daily. This was done for seventeen days, before labour came on. She had a very critical confinement. But to the patient's great disappointment, her previous sufferings and inconvenience continued as before. She now came to the writer "as a last hope." She had all the ordinary symptoms of stone in the bladder, such as constant desire to urinate, which was worse *after* the act ; had great pain also during locomotion. The urine at first was cloudy, but afterwards contained blood and mucus in large quantities— which increased as she became worse. In addition to the above and other ordinary symptoms of stone, she had supplementary symptoms peculiar to her own case. "I have (she said) been worse for two years, and have been unable to lie on my left side. I have not been able to stoop the whole time without excruciating agony—if I went on my knees, I could not get up without help. I have tried to raise myself by the help of a chair, but failed to do so, and was at last obliged to ask my husband to pull me up."

After examination a large stone was found in the bladder, measuring respectively in three directions twenty-five, thirty-six, and forty-seven millemetres (about two-and-a-half inches) when seized by the lithotrite. The patient was at once sent to the writer's establishment in Clapham. Satisfactory efforts were made to improve the inflamed condition of the bladder by the spray treatment, and this improved also her general health and thus greatly contributed to the very successful issue of the case. The success met with in other cases gave the writer the fullest confidence that his treatment would in this case bring the patient's bladder into a state tolerant of instrumental interference, and in fact a rapid subsidence of the inflammatory symptoms followed. The case fully exemplified the superiority of the writer's method of treatment of stone in respect of the following auxiliary advantages. The extreme morbid sensitiveness witnessed in stone cases is perfectly commanded and kept in abeyance; renewed life (so to speak) is given to the patient; the subsidence of cystitis gives sleep,—hence tranquil nights; the patient awakes in the morning refreshed and without the peevishness and fevered tongue usual in some cases; the stomach regains tone; the alvine secretions (from the bowels) assume a healthful condition and become abundant; and the patient feels satisfied with himself and with the physician. In short, the patient is conscious that improvement is taking place. The fact is, that more is done in these cases by the writer's *preparatory treatment* than has been done before, hence the patient's hope and confidence as to ultimate recovery is established, and this alone is an influence greatly helpful to both patient and

L

physician. When the patient in the present case was in a fair condition for undergoing the operation, she was removed to the writer's Home Hospital, 10, Dean Street, Soho, to be more within reach of his personal attention. On the 29th June, 1879, she was placed under the influence of ether, and in the presence of five medical gentlemen, and the patient's sister, the stone was thoroughly removed in one hour and thirty-five minutes, nearly half an hour of which was occupied in rectifying unexpected difficulties which had arisen, and placing the patient comfortably in bed. The stone was seized in its longest diameter. It was so long that the lithotrite could not be locked. Mr. BANKS (from the firm of Messrs. MAW, SON & THOMPSON) was present to witness the behaviour of the instrument, which had been manufactured expressly for the occasion. He was requested to test the strength of his manufacture by tapping it with a mallet, which he did with considerable force, as well as confidence. The stone was now sufficiently reduced to come within the grasp and lock of the lithotrite. Being thus reduced into smaller fragments, the whole of the calculus was pulverized to be brought away through the evacuating apparatus, and this was done without difficulty.

The *débris* on being weighed was found to be 1320 grains, or two ounces and three quarters. There were scarcely sixty drops of blood lost during the whole operation. The patient made a rapid recovery and left the Home Hospital in ten days perfectly well. Scarcely a grain of *débris* was passed after the operation. The abundant mucus and blood which the patient had been passing the whole of the time she had been ill rapidly disappeared, and the urine soon became clear and abundant— thus proving the absence of further mischief.

Before the patient left the Home Hospital she was well examined lest another stone or any *débris* had been left behind to form a nucleus for other trouble. Nothing was, however, found, and the patient has continued in perfect health up to this date (April, 1882).

No. 35.

This case like the preceding had been unsuccessfully treated by practitioners of both schools. The disease was due to uterine mischief.

Mrs. M. C., aged 33.

This case was very peculiar, and deserves special and extended extracts from the writer's notes. The patient's experience during many years of suffering (as stated by herself) may usefully be summarized, in the first instance ; the application of the treatment which effected her cure may then be described and a few general observations added—observations suggested by certain pathological features which the case presented.

Suffering from irritation of the bladder she had gradually become worse, and down to the year 1876, three confinements (with still-born children), six miscarriages, and one healthy living born child, characterized her married life. In 1872 she became pregnant, which condition made her bladder disease more troublesome than usual, and a homœopathic physician was consulted. His treatment of her case as one of " coming down of the womb," being without benefit, he ultimately recommended change into the country. While there (at Middlesbro', in Yorkshire), Mrs. C. was seized very suddenly just before dinner with unusually severe symptoms, accompanied by " dreadful, awful torture," constant desire to pass water with intense forcing pain, extending to the back passage, hips, back and groins,—pain "ten times worse than that of child-bearing." Dr. P. of Middlesbro' attended her for " acute inflammation of the bladder ; " but six weeks treatment not affording relief, she returned to London for further advice. The journey was exceedingly trying and fatiguing, the constant desire to urinate being attended with spasms so severe that she nearly fainted several times while in the train. On reaching her residence she sent for her usual medical attendant, who visited her and prescribed hot fomentations and opiates, without benefit. He consoled her however by saying that he hoped she would be better after her confinement, which was expected in five months. After five long months of terrible torture she was delivered of a still-born child. Her bladder symptoms however were in no way better, but a great deal worse, and laudanum injections were essential in order to obtain ease and a little sleep.* After three months further unsuccessful treatment he told her he did not know what more to do for her, did not understand the nature of her complaint, and had never during the whole of his experience met with a case so persistently unyielding to the remedies he had been using. Expressing great regret, he urged recourse to some hospital physician who had made a special study of female diseases. The late Dr. P., obstetric physician of Guy's Hospital was accordingly consulted. He pronounced the cause of all her discomfort to be " misplacement of the womb," and recommended an instrument (a pessary), which was introduced and worn, but gave no relief, and in December, 1872, her sufferings became intolerable. In January, 1873, another homœopathic physician was consulted. By his advice the instrument she was wearing was removed, and rightly as will be seen from the sequel ; but after treatment for five months he also expressed regret that he could do no more for her. She describes her sufferings at this time as "continued torture." She says, " night and day the desire to pass water was repeated every few minutes, attended by cutting, burning, urging pains." She also adds, " my eyes, during these terrible paroxysms, seemed to be violently forced out of their sockets ; my back and hip bones felt as if they were being wrenched asunder, forcibly reminding me of the grinding pains of labour. To bear the pains during these severe spasmodic attacks, I found relief by pressing the knuckles of both hands on any hard substance, as a table : this I repeated so often that corns were formed on my knuckle-joints."

* Mrs. C. informed the writer that her suffering was so great that she took chlorodyne for a year-and-a-half, and for the last six months before she saw him she had taken 60 drops three times a day.

The next visit was to the Hospital for Women, Soho Square. The physician in attendance treated her for " falling of the womb," and applied a pessary, saying, in reply to a doubt expressed by the patient, that " the pain she had in urinating and the frequency were occasioned by the womb pressing on the neck of the bladder ; and that as the womb went into its proper position, the bladder symptons would get well." The patient attended the hospital regularly until the day before it closed for repairs, in September, 1873. It was at this juncture that Mrs. C., then residing at Lewisham, had personally to transact a matter of business away from home, and not being able to go in any kind of carriage, as shaking increased her suffering, she "crawled out," but was obliged every few minutes to enter shops, begging permission to ease herself. At one of these, the shop-keeper (Mrs. H.), seeing her distressed condition, and sympathising with her, strongly urged her to consult the writer, and at the same time handed her a copy of his phamphlet on " diseases of the bladder, cured by a new discovery." This resulted in her seeking his aid. When Mrs. C. consulted him, though only thirty-three she appeared at least ten years older, and as usual in such cases her body inclined forward to relax the muscles of the abdomen, and thus relieve pressure on the bladder. This condition, as well as the oft-repeated doses of thirty to eighty drops of laudanum which she had been in the habit of taking, accounted for her sallow, haggard and worn-out appearance.

The case required a succession of operations which were always conducted under the influence of anæsthetics. Improvement began after the fourth application ; after the fifth the "awful pain" as she described it entirely left her, though frequent desire to micturate still continued both day and night for a considerable time ; but this the patient " did not mind so long as the *pain* did not return." The improvement was only very gradual, yet in four months she had occasion to pass water three times only during the night ; during the day no inconvenience was suffered, and at the end of five months the case was cured ; but for precaution, treatment was continued another month.

Her husband on returning home from a long voyage found her, contrary to expectation, quite well. When subsequently signs of pregnancy presented themselves some of her old symptoms threatened to return, but no severe pain ; the comparatively slight irritation felt being attributable no doubt to her pregnant state. In due course she gave birth to a full-time healthy child. Some months after this bladder symptoms again caused a little uneasiness, which was effectually removed by the application of a few more sprays, and in November, 1876, she was .apparently perfectly cured.

The peculiarities of Mrs. C.'s case call for special observation. All the medical gentlemen (one excepted) consulted on her case previously to the application by the writer of the treatment which effected the cure, concurred in the opinion that her bladder disease was due to " falling of the womb." Now it is a well-known fact that usually when " falling of the womb " causes pressure, and the woman becomes pregnant, the womb ascends from the pelvis (the bony cavity in which the womb and bladder are placed), thus removing both the pressure and its attendant inconvenience. In the present instance, however, the patient became *worse* as gestation advanced, the instruments introduced (the pessaries) doing no good, but positive harm. It is astonishing how many of these patients present themselves with all sorts of clumsy mechanical contrivances, showing how various have been the unsuccessful attempts to effect a cure. The constant straining bearing down pains experienced in bladder diseases dislocate the womb ; in other words the uterine misplacement, " falling of the womb," is the *effect* not the *cause* of the bladder mischief. It is not intended by this to convey the idea that "falling of the womb," does not *occasionally* produce bladder inconvenience : but the *majority* of bladder diseases in women are not attributable to that cause (falling of the womb). Mrs. C.'s case well illustrates this. Previous to her cure, nine pregnancies resulted in three still-born children and six miscarriages. After her bladder disease was cured she gave birth to a full-time healthy child—the object of her heart's desire. Explanation of all this is found in the anatomy and physiology of the uterine system. The same plexus of nerves, " hypogastric plexus," supplies the uterine and vesical organs with nervous power : it likewise supplies the rectum (the lower bowel) : hence the sympathy existing between all these organs; hence too, when impregnation takes place women commonly enough experience irritation of the bladder sympathetically ; but this is quite independent of any mechanical pressure caused by the enlarging uterus, inasmuch as this kind of bladder irritation occurs during the first few weeks of pregnancy, while the womb is as yet too small to produce irritation of the bladder through pressure. Certainly a converse sympathetic condition sometimes exists through pregnant women inadvertently taking strong aperient medicine ; in this case the irritant acts on the rectal branch of the hypogastric plexus, and the irritation set up is thence conveyed to the gravid uterus through the uterine branch of the same plexus, and premature labour is often-times the consequence. Many married women, seemingly barren, have borne children after the removal of a slight bladder disturbance and this in cases where uterine treatment had proved of no avail.*   The truth is that comparatively little is known by the medical profession respecting bladder pathology ; and hence bladder diseases in both sexes remain uncured, and scarcely to any appreciable extent relieved by ordinary treatment. If physicians more generally made *one disease* only their special study, something like accuracy of diagnosis in special diseases would be attained, and suitable treatment applied.

---

* The writer had the good fortune to treat a case of this kind successfully after three obstetric celebrities in England, France, and Scotland had unsuccessfully treated the cause of the barrenness as uterine.

Notwithstanding the writer's large experience and practice in connexion with the peculiar and intricate diseases of the bladder, and the successful application of the treatment,—to the discovery of which he was led by careful study,—he feels, as mentioned in the preface, that more time, systematic study and careful practice, are still needed before he can adequately present the subject or his mode of treatment to the profession generally. In Mrs. C.'s case, the patient was under the personal care of the writer for six months, and for some time he despaired of curing her. The patient's statement, " I felt I could not hold my water another moment; my desire came on suddenly, just as I was going to take tea," reminded the writer that the descent of stone from the ureter into the bladder, generally comes on precisely as the patient described her case; her disease, however was not stone. But careful observation, steady perseverance and special noting and study of each peculiarity and minutia, gave the key-note. The case had been one of long duration severe symptoms apparently only coming on suddenly. This is frequently the case in many dangerous diseases and derangements of vital functions. The suddenness of the disease is sometimes to the bladder what the " last feather" is to the " camel's back." It is so in acute pleurisy which not unfrequently ushers in an attack of pulmonary consumption. It is so with active hœmorrhage from the lungs (spitting of blood) ; the tubercular mischief, perhaps of years' duration is the *cause.* It is quite an error to suppose that spitting of blood come on primarily and consumption secondarily. It is true no very marked symptoms of consumption evince themselves before spitting of blood sets in ; but the disease has been dormant perhaps for years. The first serious *visible* symptom is spitting of blood which is simply the effect,—not the cause. It is so with disease of the prostate gland in the male. The sudden acute pain attended with retention of urine is only the *effect* of a long-standing enlargement of the prostate gland.

No. 36.

Villous tumour and stone in the bladder.—The patient was in a dying condition when brought under the writer's notice.

MRS. S., (written by herself for publication).

[Copy—Names only being omitted].

"June 26, 1874.

" In consequence of severe domestic affliction in January, 1871, I was constantly exposed to changes of atmosphere such as rushing from heated rooms to the external air which gave me severe colds and discomfort in the bladder. This at length became very severe and caused irritation of the bladder; the desire to micturate soon became constant. I sent for our family doctor, Dr. N., who gave me some medicine saying I would soon get better. At the end of three weeks I became much worse and my doctor told me my disease had turned into inflammation of the bladder. I was now very ill, and could not empty my bladder at all in the natural way, but had to stand in the erect posture, and then suffered excruciating cutting pain. I was ordered more medicine, linseed tea, and twelve leeches, which gave me some relief. At the end of a fortnight I became worse : the irritation and desire to pass water were now intolerable ; twelve more leeches were applied, but receiving no benefit the doctor made a further examination, and found (as he said) that the cause of my trouble was the womb pressing on the neck of my bladder, and intimated that as he had found the cause I should soon be relieved. I was ordered to wear a pessary which increased my suffering tenfold. I was now quite confined to my bed and applied hot sponges, laudanum poultices, and took morphia. In March of the same year my symptoms increased in severity, when a physician was called in consultation.

I was examined without success for stone. Both the physician in consultation and the family doctor concurred in the opinion that my disease was chronic inflammation of the bladder, the result of previous acute inflammation. A fresh form of misery now presented itself. Spasmodic attacks of retention of urine lasting sometimes for several hours seized me at repeated intervals. I was often compelled to take as many as six warm baths in one night; the baths and repeated doses of morphia were necessary to give me sleep. This state of things continued till the middle of April, when I made a superhuman effort to dress myself and was conveyed in an invalid carriage by rail to Hull to be under the personal care of the same physician who had seen me before. He saw me daily for ten weeks. I became worse after reaching Hull; the pain and constant urging to urinate were more severe than ever; I could not remain in bed, but had to sit in an arm chair propped up with my knees drawn up, from nine in the evening till two o'clock next morning. 1 was in this condition for ten weeks, during which period I was examined for stone on several occasions. My physician was as kind to me as any parent could have been, and regretted much he could do me no more good. He advised however that Sir H. C. should be invited to a consultation. My bladder was again examined. Sir H. C. confirmed the physician's opinion, and said I was suffering from chronic inflammation of the bladder. I was grievously disappointed when I was told that nothing more could be done for me. I was recommended to return home, and was told that as the warm weather was approaching (it was now the end of June) I might receive benefit from the genial breezes of summer. I returned to Bridlington Quay, and my husband who was just as disappointed as I was, sent for another doctor (Dr. H.), who went over the same ground as the other doctors had done. He called my disease a reflex irritation arising from the brain and spine. He tried a change of medicine and injected a solution of nitrate of silver and occasional leeching. The injection made me ten times worse, and I now was left in a most pitiable condition. My physicians resigned my case to nature. I believe nature did more than the doctors. I got well enough towards the end of August to go to London. I remained with some of my husband's friends in the vicinity of

Wandsworth. Mr. S. J., of St. Thomas's Hospital was recommended to me as a very clever surgeon in diseases of the bladder. I sent for him and was under his care more than two months. He made repeated examinations of the bladder for stone or any other existing cause to account for my continued suffering. I took acid medicines to correct my urine which was (he said) in a highly phosphatic condition. Under this treatment my general health somewhat improved; but as Mr. S. J. could not discover the *cause* of my disorder he took a final leave of me, hoping it would wear off in time. I returned home in November to my sick husband, who had become worse during my absence. I did all I could to conceal my own miserable state so as not to depress him. The family doctor again took me in hand but did me no good. From December, 1871, to 1873 I dragged on a most miserable weary life of unceasing pain; nothing that was done for me gave the slightest relief.

" I was recommended at this period to try what change would do again. My brother, living in the vicinity of Manchester, invited me there, with a view of seeing a medical gentleman of great repute, who had only recently come from Guy's Hospital. Having put myself under his treatment, I had to undergo a succession of examinations and applications of a most painful nature, which gave me more pain than all previously had done. He could discover no cause for my great distress, and gave me nothing that afforded any relief. I was again dissatisfied and disappointed, and told him so. The doctor, feeling full confidence in his own infallibility, was angry with me, and said there was not much the matter with me except nervous fancies. So much for his mature judgment! My next step was to Liverpool, to the medical celebrities there. I was again examined and again disappointed: I got no better. I was well nigh tired of doctors, and of my own wretched bodily and mental condition, and returned home in February, 1874, completely worn out with repeated examinations and failures. I was now quite exhausted and bewildered, and looked upon my case as hopeless. I gradually got worse : I was only able to crawl from one room to another. For four years my suffering was such that I had not been able to sit with any comfort in a chair for ten minutes together. I had not sufficient strength to stand in an erect posture : I was obliged to lie in a recumbent position, till the spasmodic urging and bearing down compelled me to make attempts to urinate, which I could only do in a small quantity at a time. I was obliged to take morphia at repeated intervals, which appeared to enfeeble my mind. My malady and the morphia together so upset me, that I could neither read nor even do a little needlework. I was comforted somewhat by the doctor saying I might be better after change of life took place. In March, through the blessed intervention of Providence, my brother, at Nafferton (on the York Wolds), sent me a ' Treatise on Diseases of the Bladder, Cured by a New Discovery,' by DAVID JONES, M.D., of 15, Welbeck Street, Cavendish Square, London. After perusing it I determined at once and contrary to advice to place myself under his care, and made arrangements accordingly. Dr. JONES informed me I had impacted stone, and a vascular tumour in the neck of the bladder. I was operated upon successfully. I was so relieved after the operation, that in a few days I was able to leave off my morphia pills (which I had taken with me), and to sleep the greater part of the night, and awake refreshed,--a pleasure I had not experienced for years. In three weeks my disease was cured, having suffered very little pain throughout. I was now able to take my seat at the table, walk out in the garden daily, and sleep soundly and naturally. I remained under Dr. JONES' care some weeks longer, to regain my strength. I owe a debt of gratitude to the doctor for his skilful and successful treatment of my case, and for his gentleness and kindness during my stay with him. The domestic arrangements of his hospital are very comfortable ; every department is well managed ; the house-surgeon was kind, attentive, and courteous to every one. I can only add, for the benefit of those who may be suffering as I was, or from any kind of internal disease, they ought to lose no time in being restored to health, which my experience proves they will be, if once under the care of Dr. JONES."

No. 37.

Tumour in the neck of the bladder, behind which was a rough and pointed calculus.

Mrs. E. A., aged 52.

It will have been seen from previous pages that the cases of bladder diseases which the writer has so successfully cured, have been called by numerous names by various physicians and surgeons consulted by the patients. Thus :—" inflammation of the bladder," "catarrh of the bladder," " nervous disease of the bladder," " ulceration of the mucous membrane of the bladder," " stone in the bladder," &c. Why should there be so many different opinions on bladder diseases ? The writer can only offer one solution, viz., because bladder diseases are not understood. If a dozen physicians were consulted on a case of pleurisy (inflammation of the pleura), pneumonia (inflammation of the lungs), or ague (intermittent fever), they would all diagnose pneumonia as inflammation of the lungs; pleurisy as inflammation of the pleura ; and ague as ague (intermittent fever) ; because these cases are of common occurrence and easy to recognize ; in other words these diseases are tolerably well understood by the majority of the medical profession, each being known by one name and that the *right* name. The reader may take it for granted that when several different names are given to any one disease, the disease in question is an obscure one or is not understood or requires *special* skill. No disease can readily be cured that is not correctly diagnosed. No class of diseases more clearly proves this than diseases of the *bladder*, *prostate*, and *urethra*. The following case is one in point.

Mrs. E. A. had been suffering from distressing irritation of the bladder for many years, which troubled her night and day, and eventually quite disabled her from attending to her domestic duties. She was under the care of an old friend Dr. ——, a homœopathic physician for a considerable time who called her disease by a variety of names. He did all in his power to afford relief but without success. Finding homœopathic remedies of no use, he suggested a consultation with an eminent physician attached to Guy's Hospital. She was sounded for stone and underwent a variety of examinations with a view of ascertaining the cause of her discomfort. All the means employed failed to give her the slightest relief. Another eminent physician (and an author of a recognised work on midwifery, and physician to two hospitals) was consulted. After long continued treatment and great attention from him she was advised to try change of air. She left Norwood for Llandudno in North Wales, where she remained a considerable time. She became worse during her stay there. All her symptoms increased in severity. After considerable difficulty and inconvenience and by the aid of various urinary contrivances she reached home. Finding that all ordinary means had failed to give relief she consulted the writer. The case was diagnosed to be a very small calculus which was impacted in the mucous membrane of the bladder. On the second visit she was placed under an anæsthetic and the calculus removed. The patient was very speedily relieved, the severe spasmodic pain which she suffered from being cured, but owing to other complications the desire to urinate continued very troublesome. On the whole the case was very obstinate, but she ultimately recovered and still continues well. After her recovery, which to her was quite unexpected, she wrote to one of her former physicians to say that Dr. JONES had cured her ; he wrote to congratulate her and ultimately paid her a visit. When he met her he greeted her and complimented her on her improved appearance, and then, taking both her hands, said, " I have come to you with three words,—' ignorance,' ' congratulation,' ' rejoicing.' "

The following letter has been received since the patient's recovery and is (with permission) published verbatim :—

[COPY.]

" Upper Norwood,

"Sept. 23, 1876.

" DEAR DR. JONES,—Your remark the other day did not escape my observation that you ' intended publishing all your successful cases of bladder disease ;' and as through your skill and attention I am amongst the happy number, I beg you will refer any one suffering to me for confirmation. I would with pleasure tell any one particulars of my case, and express with all earnestness the gratitude I feel towards you, as well as the high opinion I have of you which might be called flattery, were I now to write my senti-ments. Trusting that my recovery may lead others suffering from so painful and distressing a disease to a consultation with you is, dear Dr. JONES, the sincere wish of

" Yours very faithfully,

" E. A."

_____

No. 38.

Disease of bladder and urethra.

Mrs. T., aged 68.

Consulted the writer on the 24th April, 1879.—Her sufferings had been of three years' duration. She had been attended by several medical practitioners, not only without cure but with little or no relief. The case had been thus treated by two local practitioners and by two eminent surgeons attached to two of our principal metropolitan hospitals. The enlightenment of the patient amounted in effect to this (using her own words) that " I had no stricture, cancer, or tumour, but they did not tell me *what* was the matter with me." When Mrs. T. consulted the writer he told her after a careful examination that he could cure her by a surgical operation. She at that time refused her consent desiring first to mention the matter to one of the surgeons previously consulted. On doing this and telling him that Dr. JONES had said he could " cure her by an operation " the reply was " well, Mrs. T., all I can say is that no *respectable* medical practitioner would do more for you than we have done at your age." This prevailed for a time, but after considerable delay the treatment pursued by two of these " respectable " medical practitioners still continuing ineffectual either for cure or for permanent relief, Mrs. T. became for a time an in-patient at Bolton House, Clapham Road. While there she was very ill and intensely nervous, and declared that she could not then undergo the treatment. It was accordingly arranged that the writer should attend her at her own home. The symptoms of her case were very severe—(1) Frequent total retention of urine, so much so that she was obliged to have it drawn four times on one occasion at very short intervals with instruments—(2) Constantly recurring desire to urinate : for instance ten times during the night, and in extreme pain—(3) Spasmodic contraction and relaxation of the sphincter situated at the neck of the bladder (a small muscle that regulates the discharge of urine).

After short preparatory treatment, a day was fixed for operation the principal object of which, as practised by the writer, is at once to remove the *cause* of the sufferings, with a minimum of danger.

The patient was placed under the influence of ether, and the operation was performed in the presence of two other medical gentlemen, the result being eminently successful.

The patient experienced very slight inconvenience and made a rapid and very satisfactory recovery.

From early in the year 1879, the patient has had no return of her former disease to the end of 1881 and during that period Mrs. T. has seen the writer twice only, once to present him with a subscription for the Free Beds of his Home Hospital, and on another occasion to be relieved of a slight cold which she was afraid might affect her bladder, but which did not do so.

No. 39.

Chronic disease of the bladder and incipient Bright's disease of the kidneys.

A. L., aged 41, spinster.

April 3rd, 1878.—The writer's notes of this case furnish the following :—Has been suffering for four years from intense irritation of the bladder, which (to use her own language) " has become worse and worse, and now it is dreadful to bear." She has been under medical treatment constantly for two years. Her suffering has been of gradual growth. She did not think much of her discomfort at first, as it only amounted to greater frequency in passing water, the act increasing however both day and night. Her symptoms gradually changed in character : burning pain soon added—ultimately " bearing down," throbbing, and indescribable discomfort. Has been under the care of several doctors with only very partial relief. For the last two years, finding medical treatment "perfectly useless " (as she said), she has done the best she could by simple means, patiently resigning herself to her sufferings. At the time when she consulted the writer she urinated every five minutes during the day, with great pain and forcing, having to get up many times during the night without

M

effect. The urinary secretion is cloudy and full of threads and clots of mucus; it is highly alkaline, and smells ammoniacal. There are also traces of albumen—the specific gravity 1016. Since the urine has become more cloudy she has complained of pain in the loins, giving evidence no doubt of the kidneys being implicated : the specific gravity and albuminous state of the urine shows this. Administered a spray, and told her to come again at the end of a week.

April 10th.—Seven days after the first application she greets the writer with a smile, adding " I am better. I have gone as long as an hour without passing water, and when I do so it is not attended with so much forcing and bearing down pain."

April 13th.—She says, " I am so much better, sir. I have gone two hours, and the pain has *quite* left me."

May 18th.—" I am quite well ; indeed, I have been well for the last fortnight. I only pass water four times a day, and once during the night. I should not have troubled you to-night only that we are going out of town, and I thought I should feel more satisfied. My affliction had made me so nervous that I cannot make up my mind that I am cured, as it is only six weeks since I first came to see you."

Having written to ask if she continued well, the following reply was received : —

" Belsize Park, July 3rd, 1878.

" DEAR SIR.—I am most happy to say that I consider myself perfectly well. I cannot find words to express my gratitude for the kindness and benefit I have received at your hands. I can *never* repay you, even if I had means to do so. Will you please sir to accept my poor, humble, but *most grateful thanks ?*

" I am, your obedient servant,   ·   " A. L."

April, 1882.—A. L. continues quite well up to this date, more than three-and-a-half years since she was cured.

Chronic and (supposed) incurable disease of the bladder and womb, unsuccessfully treated by allopathic and homœopathic means.

Mrs. E. L., aged 35 years.

This patient consulted the writer in the year 1878. She had been suffering for eight years, and although under medical treatment for the greater part of that time, both allopathic and homœopathic yet without cure or even permanent benefit. In the subjoined letter (which somewhat fully describes her case) she says "It is only since I have placed myself under your care that I know the happiness of being free from suffering," adding in another part of the same letter, "I only regret that I should so long have allowed the opinion of prejudiced professional and other persons to prevent me consulting you." Her symptoms were of the most aggravated kind—constant urination night and day—severe burning spasmodic attacks lasting for a lengthened period, attended with bearing down pain attributable to uterine complication. The treatment employed was purely local and mechanical, and effected a perfect cure in six weeks, thus making it manifest that the patient's continual sufferings whilst under the treatment of her former medical attendants during the preceding eight years, must be attributed either to inaccurate diagnosis, or inappropriate treatment of her case.

The subjoined letter was subsequently received from the patient with full liberty (as will be seen) to publish the same.

During a tour through Hampshire in the summer of 1881 the writer called on Mrs. L. and found her perfectly well. In answer to questions relating to her health she replied "I am quite well and have not had an ache or pain since you cured me. I got over my confinement without any return of former symptoms."

Southampton, 19th December, 1878.

"DEAR DR. JONES,—I sincerely hope that my long silence has not caused you to think that I have so very soon forgotten " Bolton House" and my associations with it during the three months I was a patient there. The gratitude I feel for the kindness and comfort experienced as well as for your skilful medical treatment cannot be expressed. My only reason for not writing before has been that I might be able when I did so to speak with greater confidence and assurance of the cure effected. I have now been home six weeks during which time I have been busily engaged with domestic duties without having had the least return of my old symptoms. I have been quite free from pain and feeling quite well. When I contrast this happy state of experience with the eight long years of almost continual pain, which when I was seized with a spasm amounted to indescribable agony ; and then think too of all the weariness, debility, inconvenience inseparable from a diseased state of the bladder, I feel deeply and intensely grateful to God who directed me to place myself under your treatment and also to *you* as the instrument in His hands for effecting so happy a result. My first illness from this dreadful malady took place in 1870—it was brought on by getting up too soon after a miscarriage and by taking a severe cold. I was under allopathic treatment for three years but experienced very little benefit ; the latter part of the time I grew worse rather than better. This being the case I determined to try homœopathy and informed my doctor of my intention. He kindly expressed a wish that I might derive benefit from the change of treatment but said that he had very little hope that it would be so ; on the contrary he feared I should never be cured. Much depressed and almost in despair (for at this time I was a fearful sufferer) I consulted Dr. C——, a clever homœopathic physician. He cheered by assuring me my case was by no means hopeless—that he considered it curable. Three months of his treatment effected more than all before ; but still the disease was far from being conquered.

Through his removal to a distance I subsequently consulted two other homœopathic physicians. I derived great benefit but only up to a certain point beyond which I never got, having always more or less pain when urinating, great debility, and occasionally those dreadful spasms already referred to.

"It is only since I have placed myself under your care that I know the happiness of being free from suffering. I quite believe that the bladder affection was perfectly cured in a month and that I should have left Bolton House in that time had it not been for the misplacement of the uterus, which you discovered and also cured.

"I shall but consider it to be a pleasure as well as a *duty* to recommend your treatment to those suffering in like manner, and shall always most willingly give my testimony to any that may be referred to me. You are quite at liberty to publish any extract you may think proper from this letter and I hope this may be of use in inducing others to consult you. I only regret that I should so long have allowed the opinion of prejudiced professional and other persons to prevent me consulting you.

"Trusting that your life may be preserved many many years to be a blessing to sufferers,

"I am, dear DR. JONES,

"Yours faithfully,

"E. L."

"P.S.—I should prefer my initials being used rather than my name in full; but you are at liberty to give my name in full to any one desiring to communicate with me.—E. L."

No. 41.

Disease of the bladder in a very elderly widow lady.

E. D., aged 77 years.

This patient consulted the writer on the 29th September, 1881. She was accompanied by a friend. From their statements (noted at the time) it appeared that Mrs. D. had during seven or eight months been under private allopathic and homœopathic treatment, and in the Bloomsbury Dispensary and the Homœopathic Hospital, Great Ormond Street,—her last ticket there being numbered 148,582, and dated August 30th, 1881. Mrs. D. said :—"None of the doctors appear to me to know what is really the matter with me, and call my disease by such different names. Some say I have a cancer, others that it is old age and that at my age I ought not to expect to get well; others, inflammation of the bladder, and as I get no relief I suppose they don't know what it is. Hot fomentations have given me more relief than all their medicine." Other notes respecting this patient show the following :—Comes from a long-lived family— her mother died at 94—has now a sister living 91 years old, and says, " I might attain a similar age if only I could get cured of my painful disease." She attributes the actual commencement of her sufferings to an accident—falling down stairs—and, when consulting the writer, described her symptoms thus :—" I pass water every half-hour during the night, but am always better in the day time, and don't pass water quite so often. I am seized with a sort of spasm, which is killing me. No one can tell how dreadful my sufferings are who have not seen me during my long suffering. The constant straining and want of sleep at. night is dreadfully trying. One would think that doctors ought to give me *some* relief. Do you think your treatment will do anything for me ? " In diagnosing the case the writer found the bladder so irritable that the smallest quantity only of urine could be retained—it being rejected, spasmodically (so to speak), as soon as secreted. The urethra also was in such an irritable condition that the catheter could only with difficulty be introduced. Treatment was applied, and a spray administered.

October 12th.—The patient said :—" I have been better, sir, I am so truly thankful that I can get a little sleep. I still pass water very often, but you have eased the spasm, and I rest between times." Similar treatment again applied.

October 24th.—The patient looked better and said :—" I am very much better thank God. You are the only doctor that has done me any good. Since I was last here I have only had to get out of bed three or four times during the night, and I pass water in larger quantities without pain, which is a great comfort to me. All that the other doctors did for me was to change my medicine, and look at my tongue." The writer this time administered a tonic spray to the patient, and requested that on the next visit she would come with her bladder as conveniently full as possible.

October 31st.—The patient said :—" I am quite well sir, and am going to run away from you to live with my daughter in Uck-field, ten miles from Tunbridge Wells. My cure is really a miracle sir ; I never expected to get well. I have kept my water as you wished me since breakfast, and could keep it still longer." Catheter introduced—twelve ounces of urine drawn, clear and natural,—all the irritability was removed,—the case was cured.

The patient, in expressing a wish to have her case published "for the benefit of others," added :—" Thank God I saw your advertisement. My daughter (EMMA DOSSETER), who lives with Mr. WOODS, Dane Hill, Uckfield, first showed it to me, knowing what a great sufferer I was."

NOTE.

The above case suggests the following observation :—The writer has found that in almost every such case coming under his notice and treatment the patients have been discouraged, both by their medical attendants and others by being told :—" Poor old soul, what can be expected at your age ? " The writer has always felt that the more infirm or aged the patient the greater the reason for *something* being done to at least ameliorate the suffering and render life tolerable—and he has had the gratification of witnessing such a result attending his treatment of many such aged persons, and moreover of *curing* bladder and other diseases in them,— thus contributing to the prolongation as well as to the comfort of their declining years.

No. 42.

S. S., widow, aged 71.

This patient consulted the writer three years ago (June, 1879). The following statement of her case was given by her daughter.

Her mother was quite well till October, 1878, when she had an attack of neuralgia and rheumatism. In February, 1879, she began to complain of pain in passing water which gradually increased in severity despite of all her doctor did for her. Her disease was called "chronic inflammation of the bladder." The inflammation was said to have run on to "ulceration of the neck of the bladder." The doctor tried everything he knew of without effect. The bladder was washed out twice a day, still the patient got worse. A month before the writer commenced his treatment a physician from Maidstone had been sent for who corroborated the opinion of the family doctor, adding "it is a thorough breakdown of nature." The physician's prescriptions were nevertheless carried out faithfully. Mrs. S. got no relief from all that had been done. Finding that the case was "incurable" the family doctor after three months constant attendance said "I can do no more for her and I will only call occasionally."

The condition of the patient when first seen by the writer was thus related also by her daughter :—she said, "My mother is so weak she cannot talk to you. I will tell you how she has been. She has had to pass water in extreme agony every half-hour to an hour—the urine is very thick and sticky and smells very offensively—once it had blood in it. We have used all kinds of disinfectants but they do no good, the smell makes us all sick.

Three or four of us have been constantly with her night and day for the last three months. She was continually wanting the chamber so that we could not leave her and we are obliged to support her while on it—she is so weak—she has had no appetite during the last three months but we have kept her up with a bottle of champagne a day. On one occasion we thought she was gone —her breath seemed to have left her body, but we brought her to by dipping a feather in brandy and putting it between her lips. The doctors said her heart was weak and gave instructions that she was never to move quickly—the vessels about the heart they said were wrong."

The writer carefully examined the case and administered treatment though with some difficulty the parts being unusually irritable. Full instructions were given how to proceed till the writer should see her again. On visiting her a second time at Tunbridge he was gratified to find her propped up in bed and able to converse pretty freely—she said in a distinct voice :— "the first spray you gave me produced beautiful sleep and comforted me for three days—it was new life to me, I had only to pass water three times all night." Another spray was administered. On paying her a third visit she said "the second spray relieved me so much that I slept nearly the whole night." On seeing her for the fourth time she said " I am so much better doctor that I really think I am cured. I was out of bed and moved about a good deal yesterday and tired myself a little. I was only out of bed once the whole night and passed water very comfortably without pain. I have lost all the pain that used to torture me fearfully." The patient although apparently cured was not really so. She was so much relieved however as to lead everyone around her to suppose that she was cured. In this case as often in similar cases there is a vascular growth in the bladder needing removal by surgical means. The patient was with little or no inconvenience removed to London—surgical treatment was applied and the growth was effectually removed.

Since then Mrs. S. has been perfectly well, and now (in her 75th year) is in perfect health and free from any bladder inconvenience. She will willingly communicate with any enquirer and confirm the above statement.

No. 43.

Reputed paralysis of the bladder of 30 years' duration with occasionally severe paroxysms, endangering life.

The Hon. L. D. E., single.

The following case bears the characteristics of romance, but the facts from a professional point of view are only too sober. The central feature is the dismal tragedy of false diagnosis.

The Hon. Miss L. D. E.——, as the daughter of a distinguished man in public life, was surrounded with all the advantages of wealth and station. Health however is a higher endowment than material means. She was reduced by physical means to a condition which the meanest could afford to pity. From the infantile age of two years the state of the bladder was such as to be incapable of any retention; and the water passing away as it dripped from the kidneys: she lived in the deplorable plight of nightly urination. A vague notion prevails among the physiologically ignorant that the capacity of urinary retention is subject to the will; and a naughty habit, therefore, was visited in after years with the extremities of scholastic discipline. The reality of the situation was eventually recognised, but the ablest and most eminent advice was unavailing, and for thirty years she passed her life in the hopelessness of what was pronounced as incurable paralysis of the bladder. Occasionally the sufferer was subject to aggravated paroxysms referred by the surgeons to inflammation, and these as years passed on grew more frequent and severe, until life itself was jeopardised. In the year just passed she chanced to

light on the fifth edition of the present work and with the hopefulness however desperate which a suffering patient rarely relinquishes,
the writer was consulted to enquire if anything could be done. The
case was one demanding preliminary treatment before an effective
examination could be made, and when the bladder had been
cleared it was conducted under the influence of ether. The result
was a distinct apprehension that the reputed paralysis of the
organ had no existence. To the writer's perception the real
source of the mischief moreover was sufficiently plain. Meeting
the mother's somewhat anxious enquiry he offered the tranquil
assurance, "Madam, I think I can cure your daughter." "Cure!"
returned the lady in astonishment, "What *can* you mean? You
know doctor that all the best surgeons of the day who have seen
my daughter have pronounced her case to be incurable. The
most eminent men whom we consulted all echoed the same thing:
that it is paralysis, and that it is incurable. I only ventured to
hope that your special skill might devise some methods of relief.
Surely (in incredulous but appealing tones) you cannot *cure*!"
The truth was the writer had discovered what other surgeons had
failed even to suspect, the presence of what he anticipated from
the first, an enormous stone. The bulk of the calculus filled the
whole cavity, the bladder was intensely hypertrophied, and grasp-
ing the substance forced it into the left side of the neck of the
organ, and thus obstructed its function, insomuch that there was
no retentive capacity, and the urine flowed away involuntarily.
Hence the former medical and surgical advisers perceiving the
effect but ignorant of the cause, pronounced it "paralysis of the
bladder," "which," they added "is incurable." The writer
thereupon told the mother of the discovery, and advised the
removal by lithotomy (the cutting operation). At the same time,
he made no concealment of the element of risk in the case. The
patient's health was delicate, and her natural strength was reduced
by social privation and physical suffering. Consent however was
given, and it was agreed that the opinion of an eminent surgeon

should be taken as to the chances of success. Further, the writer's view was confirmed, on a second examination, by two surgeons of distinction. They agreed that lithotomy was the only available resource, since the lithotrite (for the crushing operation), owing to a difficulty of entrance to the bladder, could not be employed. The patient having elected a sojourn at the writer's residence, and day and night nurses appointed, the operation was successfully performed by Dr. GORDON G. JONES (the writer's son); and, not without considerable difficulty, the stone was removed. It was found to weigh over three ounces. The severe injury occasioned by it to the neck of the bladder was a complication; but the patient nevertheless made a good though a slow recovery.

Thus is illustrated the prolonged mischief of a wrong diagnosis to which even "eminent authorities" are liable. The writer forbears to speculate what might have happened if the positions of the homœopath and the allopath respectively had been reversed. If disastrously wrong treatment had been pursued for over thirty years by sundry homœopaths, and at the last moment an allopath on being appealed to had discovered it, how the whole social fabric had been convulsed with indignation! "Actions-at-law," "Malapraxix," and even "the gaol," on many an impassioned lip! The world, however, is just at heart, save only when it happens to be prejudiced.

NOTE.

The above cases are typical of bladder affections such as frequently present themselves to the physician,—each case having its own individual peculiarities,—but to which down to the present time no plan of treatment other than that adopted by the writer has given more than temporary relief. The majority of the cases treated by the writer have been radically cured in from three to six weeks. Some have occupied nine weeks. When the symptoms do not yield within the ordinary time the case is regarded as very stubborn, and although the writer does not give up such cases as beyond cure he thinks it desirable to mention such possible stubbornness in order that patients may not be disappointed if the disease does not so speedily respond to treatment as desired. Usually when the case proves difficult to cure there is some organic complication such as " Bright's disease," or some other disease of the kidneys, liver, or lungs. Impacted stone in the kidneys or bladder is another complication very difficult to manage. Cancer of the same organs also very seriously interferes with and retards the treatment. But be the cause what it may relief at least is obtained by the writer's method of treatment and far more effectually than by the administration of opiates or chlorodyne, whether administered medicinally or by subcutaneous injection, or by suppositories. Disease of the bladder is said to be far more common in the male than in the female, owing to the more complicated nature of the genito urinary organs in the male sex. Although prostatic disease enters very largely into the cause of many of the symptoms of urinary trouble and stone in the bladder is more common in the male than in the female the writer's experience nevertheless leads him to the conclusion that the proportion in both sexes is after all nearly equal. For while the prostate plays an important part in producing urinary troubles in the male, the womb and ovaries produce a great number of urinary troubles in the female arising mostly from mechanical causes such as tumours, uterine misplacement or disturbance of the nerves which supply those parts, curable by removing the mechanical causes or treating the nervous origin— as will be seen by reference to the cases of women reported in this edition.

Considering the frequency with which diseases of the bladder occur in women it is somewhat surprising that no medical or surgical writer has heretofore drawn special attention to the subject. There are numerous works on "diseases of women," and of recent date too in England, France, Germany, and America, but disease of the bladder in the female has been passed over with very few remarks. No attempt appears to have been made to classify them as has been done with diseases of the bladder in the male sex, on which subject plenty of books have been written.

No. 44.

Obscure presence of stone in the bladder discovered by means of the spray treatment, after baffling the search of several notable surgeons, Mr. Berkeley Hill and Mr. Christopher Heath (senior surgeons to University College Hospital), and Mr. Barker (surgeon to the same institution).

Mr. DAVID BOWTLE, of 36, Sidney Street, York Road, King's Cross.

To quote the patient's own account, as recorded in the writer's case book : —

"When I move about much, or go to pass water, or stoop to lace my boots, my sensation is as if a needle were being thrust into the neck of my bladder. During a long illness of four years I have been attended on and off by my local doctor who sounded me for stone but could find none. As I continued to get worse he sent me for further advice to University College Hospital. Going there on the 11th October, 1888, and being seen by Mr. BARKER, one of the surgeons, I received the same opinion ; and in like manner he examined me for stone but without finding any. To afford further opportunity for sounding me, he advised me to enter as indoor patient, and I was admitted next day and sent to ward No. 1. Mr. CHRISTOPHER HEATH, Mr. BARKER's senior, having examined me in his turn, likewise thought I had stone, but I do not think he found it for he did not say so. On the 15th

October I was further examined with the same result as before. Though my local doctor and the hospital surgeon all thought I had stone, yet none of them were able to find it. After examination on this occasion Mr. CHRISTOPHER HEATH said however, ' I will put you under ether one day and crush it for you.' Saturday, the 17th October being fixed for the operation, Mr. BERKELEY HILL and Mr. CHRISTOPHER HEATH, both senior surgeons to the hospital, were present with the doctor who gave me ether. Two days afterwards Mr. CHRISTOPHER HEATH said to me, ' We examined your bladder thoroughly ; I examined you myself for a quarter of an hour but could find no stone ; I do not think in any case it is larger than a pea, and it will come away without an operation if you go on your hands and knees and strain when you are passing water.' Glad to hear such good news and afraid of an operation when so many doctors had failed to find the suspected stone. I willingly consented to follow Mr. HEATH's advice. Accordingly I strained violently at great inconvenience and increased suffering, but no stone came. Finding after fair trial that no one either found stone or gave me relief I came to you."

After hearing the candid statement of Mr. BOWTLE, the writer requested to see him pass his water. The test was sufficient. He thereupon told him emphatically that he was *quite sure* he had stone ; but if he examined him in his present condition he would be unsuccessful like all the former surgeons in finding it, inasmuch as the stone was covered with thick mucus and blood, or possibly embedded in the folds of the bladder and could not be discovered until after an application of the spray treatment for clearing the bladder, the only known means of revealing stone under the circumstances. Here it may be discerned is an explanation of the fact that surgeons, otherwise able, fail in the case of patients suffering from bladder disease, and, prejudiced against means which they do not understand, endeavour to pacify them with the assurance (as for example to Mr. H. J. BARRETT, of Hull)* that they

---

*See case No. 45 in this edition.

are suffering from an incurable malady namely enlargement of the prostate. To resume : When in such cases as Mr. POWTLE's the urine is collected after several urinations, it resembles the sputum (expectoration) of sufferers in the last stage of consumption, which is often called "oyster-like expectoration." Hence from the condition of the bladder the concealment of the stone and the necessity for the spray.

The patient entered the Home Hospital on the 6th November, 1888. The administration of the first spray brought away nearly a wine-glass full of thick clotted mucus of such a consistency that no one could be surprised at not finding a stone in it, however prolonged the examination. A few more sprays served to clean the bladder and the urine also, and at the same time relieved the patient from the suffering he had endured so long; indeed, the complexion of the case was materially altered. The "burning, pricking, cutting pain" he had hitherto experienced passed away ; and, from being highly nervous and irritable, through constant anguish and repeated straining, he became cooler in temper, comfortable in aspect, and happy in his existence. The manager and inmates generally were not slow to remark on the change : " How much more cheerful Mr. BOWTLE looks, and how his face, has brightened up. He has lost the worn expression he had when he came !" They began to think, too, rather precipitately, that "the doctor would cure him without an operation." In truth, the patient became well enough to take a three days' interval to go home and look after his neglected business, while under preparatory treatment for the operation which was in fact contemplated.

Nor is this an unusual incident. Not infrequently, patients under similar circumstances, have found themselves so well under the writer's preliminary course, as to lead them to believe no stone was present. The cases of E. B., I. B., and I. C. W. I. may be cited as examples.*

---

*See Cases 7, 8, & 9, in the present edition of "Diseases of the Bladder," &c.

When the bladder had been well cleared, and the mucous membrane recovered from its inflamed and ulcerated state, a day was fixed, the patient was placed under ether, and the operation was satisfactorily performed by the surgeon to the Home (Dr. GORDON G. JONES, the writer's son), to which cases, of course, he is well accustomed; he quickly found the stone and promptly removed it. Thus, without hitch or hindrance, was easily accomplished what other surgeons, eminent though they be, had regarded with vain perplexity. The removal of the fragments, in the form of lithic acid, occupied no more than twenty-five minutes; their weight, when dry, was over half-an-ounce.

Mr. BOWTLE made a satisfactory recovery but was not ready to leave the Home till the 7th of December. His unusually long stay—a whole calendar month—was due to the serious injury to the bladder occasioned by the long residence of the calculus, a period of four years. He is however well pleased with the result and has since been actively and contentedly engaged in his business. He speaks highly of the spray treatment, and with good reason is convinced of its advantages, only regretting that he did not know of it years before.

- - - - -

No. 45.

Enlargement of the prostate pronounced "incurable" by three surgeons of repute—stone removed—said not to exist.

H. J. B.

This patient had been unsuccessfully treated by three well known surgeons, and was told (as many others have been), he was suffering from " enlarged prostate, from which he was not to expect any relief." " One and all declared " he had " no stone."

These adverse opinions naturally made the patient unhappy about himself. Casting his eye—as he expressed it—one day over a newspaper, he saw an advertisement of the " writer's

N

book" and consulted him. The patient was told his doctors had diagnosed his disease as enlarged prostate correctly enough, but were wrong in saying it was "incurable."

The "spray" treatment was administered with decisive effect; frequency of urination and pain consequent upon it, were speedily relieved, and the urine became clear. In other words, large quantities of mucus (the results of inflammation) was removed, and inflammation of the mucous membrane was cured. During the course of treatment the writer suspected stone in the bladder, which opinion was verified by an examination, and the stone was subsequently removed. The patient made a rapid recovery, and was soon out of doors. This case resembles the cases of Admiral Sir GEORGE ELLIOT, K.C.B., Captain A. C. CLARK, R.N., of Bombay, and many others that might be named.* These cases go to prove two things : first, defect of the means ordinarily employed to discover stone in the bladder; second, the errors of physicians and surgeons who say that disease of the prostate is "incurable." When the patient was so far relieved and comfortable as to know that enlarged prostate was not "incurable," he stated in one of his letters,—" I consulted you afterwards for a chronic functional disorder—constipation, for which physicians in my experience have found no permanent cure. My present experience is that I have better health than I have had for the last ten years ; I have no return of the pains and distressing symptoms I had before you took me in hand. I have no bladder troubles and have had no occasion to take a dose of medicine for constipation since I left you—this is an experience unknown to me for twenty years." In a letter dated 12th May, 1890, he writes, " I have been frequently in London lately, and several times proposed calling upon you, not that I require any professional advice,

*See cases I. and II. in second edition of writer's book on urinary diseases, Simpkins and Marshall, Paternoster Row, and Mitchell & Co,, Red Lion Court, Fleet Street,

but merely to report to you that as it is now about two years since I placed myself under your treatment, for disease of the prostate gland and stone in the bladder, and that I have had *no return whatever* of my troubles in this way.

" You are at liberty to make use of me in any way you like as a reference, as the thorough manner in which you did your work is tested by my past two years' experience. Hoping you will be spared many years to be a benefactor to suffering humanity, I am, my dear sir, Yours faithfully,

HENRY J. BARRETT.

No. 46.

Urethritis, spermatorrhœa, and tumour in the bladder.

O. A. R.

[COPY OF LETTER.]

"BRISTOL,

"October 11th, 1889.

" DEAR SIR,

" I have reason to fear that I have by my long silence deserved to be placed on the list of ungrateful patients·

" And I think the ingratitude of patients must be one of the hardest things a doctor has to bear.

" Yet. I trust I am not ungrateful for your kindness, and I assuredly do not forget that to you I owe my cure and my life.

" You will remember the wretched state I was in when in the summer of 1886 I entered your " Home Hospital " reduced almost to a skeleton by urethritis, spermatorrhœa and tumour in the bladder. That I had been treated for a long time but only to grow worse ; by four other English doctors and two medical gentlemen of high standing in Spain ; and how with the application of your "spray" treatment the work of restoration commenced.

N2

Under your treatment large quantities of black fœtid pus and blood were passed with some pain but without any surgical operation from the bladder.

" Now I am thankful to say that increased vitality and strength, though accumulated gradually, warrant the confident anticipation that I shall soon have the complete restoration I have so much longed for.

" When in a better position, I hope to seek you out a second time, to thank you personally for all that you have done for me.

" In the meantime, please, in the interest of poor sufferers, consider yourself at liberty to make any use whatever of this letter, reserving only for private reference the name and address.

" I shall be most happy to write out at any time, for any inquirer, as concisely and correctly as I can, a synopsis of my case and the history of its treatment, and shall always consider it a privilege to attest the genuineness of my wonderful cure, and to answer private inquirers recommended by you.

I desire to offer to your son, Dr. GORDON JONES, for his unremitting kindness and attention to my case, my sincerest thanks and deepest regards. Accept, dear sir, this slight expression of gratitude, already too long delayed, from one of the many whom you have restored to life, and believe me, ever devotedly yours.

"O. A. R."

THE SPRAY TREATMENT IN VETERINARY SURGERY.

While the writer has no wish to encroach on the domain of veterinary surgery, he is yet desirious of proving the universality of his " spray treatment " in diseases of the bladder, and publishes the following as a case in point.

A valuable and favourite cob, the property of C. G. DUFF, Esq., of 1, Lennox Gardens, had been suffering for a long time with disease of the bladder, occasioning constant and painful urination. When in harness, or at exercise, the animal would suddenly stop and strain violently, his body at the same time being convulsively distorted; this was attended with groaning and occasionally sweating. His head would be turned anxiously towards the flank, indicating that the pain came from that region. In addition to the above symptoms, he was frequently seized with a paroxysm, when he would shift from side to side in his box kicking violently; which symptoms, veterinary surgeons inform the writer, are evidence of acute pain in the bladder. The owner sought the advice of an experienced veterinary surgeon in his neighbourhood, who closely attended upon the patient for some time; but the poor animal deriving little or no benefit, a consultation was suggested with an eminent surgeon, a Fellow of the Royal College of Veterinary Surgeons. For nearly an hour an attempt was made to introduce a catheter, but the instrument failing to reach the bladder, owing to (as was supposed) "thickening of the urethral canal," further attempt was declared useless. The animal was declared to be incurable, and the owner advised to get rid of it. The veterinary surgeon who first attended the cob, being acquainted with the writer, and knowing his success in bladder cases, said to be "incurable," in the human species, called upon him and asked if he would undertake the treatment of such a case. He gladly did so, feeling assured that the "spray" treatment would meet it. An examination of the urinary secretion was first instituted, which was found to be alkaline in character, and contained mucus to the amount of 33 per cent., very thick and highly adhesive and offensive, clearly showing severe cystitis. After consultation on the case, a plan of treatment was laid down and carried out: The administration of a few "sprays" speedily restored the poor suffering cob to the owner, perfectly cured.

The writer does not claim the above cure to be due entirely to himself; indeed, it is but just to say that, but for the useful mechanical hints of his son, in the preparation of the necessary instruments, and the valuable and intelligent co-operation of the surgeon in attendance, it is very doubtful if he would have been able to carry out the treatment; all admit however, that but for the " spray " treatment the cure would not have been effected.

The owner has willingly consented to verify the above statement, as will be seen by the following letter, written seven months after the treatment.

1, LENNOX GARDENS,

May 12th, 1890.

DEAR DR. JONES,

Please make any use you like of my name as a reference. Your cure of my cob was *perfect,* and I am most grateful to you.

Yours truly,

C. G. DUFF.

MEMORANDUM.

It may be useful to the general reader to mention that the specific title of this work :—"stone and diseases of the bladder and prostate " does not by any means express all the forms of disease ofttimes associated with or resulting from the diseases thus specified, and coming within the cognizance of the specialist.

Many diseases having ordinarily certain distinctive names, and likely therefore to be regarded as not having a genito urinary origin, may notwithstanding have such an origin. A long list might be given—a few only can be here mentioned. For instance,—(1) In *young persons* :—diseases of the nervous system, due to early indiscretion, loss of nerve power, inducing depression of spirits, and not unfrequently epilepsy, the treatment of which, if applied without reference to its originating cause, is little likely to be successful. (2) In *adults* :—dyspepsia, irritability of disposition—softening of the brain and spinal cord—general paralysis—disease of the kidneys, not unfrequently associated with and treated for "Bright's disease of the kidneys," the real cause being, probably, some chronic urethral discharge or stricture. Affections of the eye, skin, throat, and lungs, also frequently supervene on a poisoned state of the blood, traceable, it may be, to syphilis or other debilitating causes.

All these, with others, may really have more or less remotely a genito urinary origin. Such, indeed, is the connection ofttimes existing between one disease and another, that amid the large number of cases presented to the specialist he frequently finds himself compelled to treat diseases which would not, perhaps, be regarded as strictly within the range of his speciality, and thus his attentions may embrace a sphere not dissimilar, in extent, to that of the general physician. The peculiar value, however, of the services of the specialist is shewn where his diagnosis reveals the *origin* of the disease, thus leading to treatment of its cause, rather than of its effects, and consequently inspiring more reasonable and better grounded hopes of success.

STRICTURE OF THE URETHRA.

Nothing is more common than for patients to say they have "stricture" when it does not exist: it is equally common for stricture to exist when the generality of medical men say there is none—even specialists make grave blunders in this respect—they introduce a good sized English bougie apparently without difficulty and tell the patient he cannot have stricture as the instrument goes in.

The writer cannot conceive a greater fallacy: he would rather take the opinion of a patient in incipient stricture than the opinion of the generality of medical men. There is great significance in what a patient himself says about stricture and he is usually right, while surgeons are frequently wrong. The patient's daily experience of his complaint has much to back it against that of a medical man. He knows he has had gonorrhœa or has committed early indiscretion, or may be has been violently kicked in the crutch, or has met with some other mechanical injury which is an occasional cause of stricture, and he sees a gleety discharge glueing up the entrance to the urethra, but he is laughed at and told he is nervous and that there is nothing the matter with him. The writer maintains that this "gleet" is the *incipient* stage of stricture. Gleet is a mucus discharge and whenever it is present it is the effect of chronic inflammation of the urethra, and although possibly it does not considerably narrow the canal, nevertheless accurate measurement will discover the commencement of contraction. Gleet is not present in a *healthy* urethra, consequently it is the effect of an *unhealthy* inflammatory state which will lead to more trouble if left to itself. When the cause of this gleet is not removed more trouble follows, and then medical men begin to see that all is not right; the patient will now tell you that the stream is smaller than it used to be; he passes water more frequently than he should, and has to force more to get rid of it and when it passes it does in two or more streams. Sometimes it splutters like water coming out of a watering pot, at other times it is flattened or passes in a spiral stream like a corkscrew.

There are many cases of incipient stricture which give no discomfort: the patient passes a good stream and the urine is to all appearances perfectly clear and normal. Such however is not the case. Let the patient pass some of his morning urine into a clean glass and hold it up to the light and he will see shreds of mucus or floating particles—flocculi—in a state of minute sub-division, contrasting greatly with *healthy* urine, which is absolutely clear and transparent. So it happens that a man may have stricture for years without knowing it, but let him examine his urine as directed above and if it contain flocculi it is incontrovertible proof of the existence of stricture. It is very necessary that this early stage should be diagnosed as soon as possible, for the earlier a stricture is treated the quicker and more certain are the chances of cure. If left to itself it will slowly and surely get worse, until at length all the troublesome symptoms of stricture will become fully developed. Persons who use injections to try to cure gleet or gonorrhœa, or to prevent gonorrhœa, suffer from worse strictures than those who do not do so—indeed injections *cause* stricture and orchitis quite independently of contagion.

In due course the stricture becomes more obstructive and more pain follows; inflammation from the urethra extends into the bladder producing cystits; the urine becomes loaded with mucus and before long the patient is unable to command his water during sleep, and it dribbles away from him unconsciously and wets the bed clothes. Not unfrequently it results in "impotency." The writer has known this advanced condition treated for all sorts of complaints: some have been told they had stone; some Bright's disease; others paralysis of the bladder, or disease of the prostate, or kidneys even. In many cases the patient's urethra had never been examined. The writer remembers seeing one case from Eccles near Manchester where the patient declared it took him (by his watch) six hours each day to empty his bladder: *i.e.*, he set aside two hours three times a day for that purpose, and then he could only pass water drop by drop with fearful straining; yet this patient had been treated for years by various medical men who never examined the canal to search for stricture but merely gave him medicine.

Stricture in many instances is of very slow growth, sometimes taking five, ten, twenty, and even thirty years before it becomes very troublesome and on this account it is overlooked and

mistaken for some other disease and only discovered after prolonged mischief has been established. If stricture does not show
itself in a few months or in a year or two, the cause of it (in most
cases gonorrhœa) is overlooked and stricture under the circumstances is said to be out of the question. By far the greater
number of cases of stricture that come under the writer's observation date their origin a considerable time back.

It has been asserted for years by surgical writers that once an
organic stricture is established there it will be for life. Take for
instance a paragraph from a modern and able writer on urinary
disease—he states, " Once acquired it cannot be dispersed by any
known means. It cannot be removed by absorption although
the contrary has often been affirmed. You may dilate it, you
may cut through it, but there it will always be. When a man
once has organic stricture he has it for ever."*

It was due to the opinions entertained by surgical writers in
England and on the Continent of the incurability of stricture that
made the writer thirty years ago determine if possible to erase
that impression from his mind and such an opprobrium from
surgical literature, and he satisfied himself at least that authorities
in those days as in the present were too dogmatic in their views
on the pathology and cure of stricture and urinary diseases in
general. The cases of stricture which he has cured by " unusual
dilatation " are so numerous as to allow of no doubt upon that
point. Thirty years' experience may not be long enough to
establish a principle, but surely it is ample time to establish
confidence in the plan above-named, and but for the length of
time which this treatment takes he would never adopt any other
method. To say the least of the treatment it negatives the
assertion made in old and modern works on surgery, as to the
incurability of stricture.

---

* " Clinical Lectures on Diseases of the Urinary Organs," by Sir HENRY
THOMPSON; London, J. & A. Churchill, New Burlington Street, 1876, p. 31.

We are indebted to Dr. Otis of New York for urging this on the profession, and irrespective of the criticism of Mr. Berkeley Hill and others; the writer mentions for the benefit of sufferers that he has cured over *a thousand cases* of organic stricture by his own plan and that of Dr. Otis's during the thirty years he has been in practice ; a very great many of these patients have been seen years after the operation, and in very few instances has the stricture evinced any signs of vitality ; in those few cases in which it did return all the patients confessed to having contracted a fresh attack of gonorrhœa.

The ordinary treatment of stricture by bougies does undoubtedly give temporary relief: but it is *only* temporary, for sooner or later the stricture will reassert itself and in a more severe form. The longer it is in existence the more aggravated does it become, it grows more cartilaginous and unyielding, and is consequently more difficult to cure.

It is a well known fact however that " continuous dilatation " if fairly carried out will enlarge the strictured urethra most effectually, and at the end of a year or two or more it will have contracted possibly a millimetre or two. The writer asserts that more prolonged continuous or *unusual* dilatation if carried on to a *greater* extent will leave the canal as nearly as possible normal.

The treatment adopted by the writer is so perfect that he knows of only one case which returned. This case was both an obstinate and obscure one, and had been operated upon three times by another surgeon. The patient however is so far cured now that it is believed he will require no further treatment.

Doubtless organic deposition in any part of the body is rarely resolved into its absolute original condition ; still this does not justify the wholesale refutation we see in surgical books as to the incurability of stricture. Once the stricture is cured it can in no way interfere with the genito-urinary functions or in any way exercise an injurious effect upon the health of the patient.

Besides the usual organic stricture there are other strictures of the urethra dependant upon a variety of causes which not unfrequently come under the notice of the specialist—such as

1. Congenital stricture.
2. Traumatic stricture.
3. Hereditary stricture.
4. The masturbator's stricture.
5. Spasmodic stricture, etc., to the treatment of which a chapter will be devoted in a future edition of the author's book on "Diseases of the bladder and prostate, and the cure of stone without cutting, &c."

ELECTROLYSIS IN URETHRAL STRICTURE.

Electrolysis is a technical expression from two Greek words, signifying a loosing or relieving by electricity, and may be popularly described as "the electric cure." Within the past few years it has received considerable development as a curative agent, and from a position of suspicion and distrust, has come to take a recognised place in medical science. Its claims are great and its character specific ; and various diseases, before regarded as incurable, are held to be cured by its aid. Its consideration is certainly of surpassing interest, and plainly demands the attention of every one concerned in the treatment of disease.' In respect of one intractable malady in particular, a recent article in an American professional publication is suggestive of thoughtfulness, "The New England Medical Monthly," of December last, inquiring "What is the present status of electrolysis in the treatment of urethral stricture," remarks that "Not long ago, physicians and surgeons of repute flouted the treatment of urethral stricture by electrolysis ; now, it is so generally and successfully practised that scarcely any one opposes it." This decisive change of opinion, it explains, "is undoubtedly due, first, to the better understanding of the electrolytic treatment as distinguished from 'galvano caustic ;' and, secondly, to the successful treatment without relapse of a large number of cases (fully reported) by many physicians of high repute. It is undeniable," the article further declares, "that the method now adopted was first grasped and put forward by Dr. ROBERT NEWMAN, of New York, who, despite misrepresentations and abuse of the ignorant, has zealously laboured for eighteen years to perfect the instruments used and the technique of the operations, until by manifest success the most sceptical are convinced." Coming to details, the success of "NEWMAN'S method" is attributed "entirely to the weak currents used, and the frequency of their application, which result in galvano chemical absorption of the diseased structure ;" while the failure of MALLEZ and TRIPPER is set down to the employment of a stronger current, "galvano caustique," involving a caustic current, and not to electrolytic action. Hence the "failure of their method, and the triumph of the NEWMAN method."

Obviously, this new curative agent commends itself strongly to the consideration of every specialist, in the particular branch of surgery indicated ; and the present writer cordially sympathises with his transatlantic brother, Dr. ROBERT NEWMAN, in the indomitable courage with which he has borne up against boorish opposition and ignorant clamour, and is ready to appreciate with candour, his magnificent efforts in the cure of stricture by electrolysis. Moreover, he has no hesitation in expressing implicit confidence in the development of the new agent for purposes of urinary surgery. He feels constrained, however, to reserve his final judgment until he has been able to realise in his own practice, and in the experience of others a wider range of cases than is yet available. Meanwhile, he will only remark, as to its application to stricture, that since this condition takes a long time, in some cases as much as thirty years, in fairly asserting itself, it is a moot question whether, after apparent cure by electrolysis, it might not recur, though in similar slow degree. On the general question something more may be said. Electrolysis is now so frequently employed in ordinary surgery, that by a steady perusal of current medical records, fair ground is afforded for a right estimate of its value. Much of the range of inquiry has been satisfactorily cleared. Dr. Apostoli has convinced the medical world that fibroid tumours of excessive proportions can be dispersed by it, and illustrative casts in this regard which the writer was enabled to see in the Doctor's Clinique in Paris, inspired him with the belief that electrolysis has a brilliant future before it. Hæmorrhage, again, of the most uncontrollable character, has been arrested and cured under its influence ; and patients thus suffering have been restored from a dying condition to hopefulness and health. Polypi in the uterus, in the rectum, in the nose, and in the ears, have given way before it ; and fistulæ and sinuses, chronic ulcers, deposits in the throat and eyes, have yielded to it. These results are well ascertained and sufficiently established. Respecting the treatment of enlarged prostate, urethral and bladder affections, the writer hopes in the near future to record more extended experiences. He has already had many very successful cases, but is afraid of being premature in offering his opinion.

A VINDICATION.

The objections of the profession—verification of cases—"professional delicacy"—publication of the writer's views and experiences—categorical formulation of his claims—their confirmation.

If the writer was unconscious of the need of authenticating his published cases, which of course he never was, there has been no lack of watchful critics to bring the deficiency home. In the earlier issues of the present work, those presented by way of illustration it is true were generally distinguished only by the initials of the patients, without mention even of their places of residence. Not from a motive of concealment on the part of the writer, but because the patients themselves were entitled to have their privacy respected. His professional adversaries, of whom he is honoured with many, were not slow to perceive their advantage. Tacitly admitting that the cases for the most part were pregnant with significance, they found occasion to point the sneer, "any one may publish wonderful cases that can be palmed off on the public with initials." Really however the writer is under no temptation to manufacture professional experiences. Those which he has published are typical selections prominent among hundreds occurring in the course of years. The implied reproach was unworthy of answer, but the public to whom the writer addresses himself, have a right to the assurance that the examples before them are something more than mythical. In more recent editions therefore permission was obtained, generally with willingness and rarely with reluctance, either for the open publication of patients' names and addresses, or else for a reference by which identity could be verified. Furthermore, in most cases the recovered patients expressed themselves perfectly ready to furnish any relevant information. An appendix at the end of this volume affords full authentication, with the facility also of direct communication. The cases published in so much detail both in the present work and in the second edition of "Urinary Diseases"* (edited by the writer's son, Dr. GORDON G. JONES) ought, in truth, to satisfy the professional critics as they evidently do the suffering portion of the public concerned. With a view, too, of obviating further cavil, the names of physicians and surgeons treating

---

* Urinary diseases : analysis of 500 cases of stone, stricture, diseases of the bladder and prostate, and obscure and supposed incurable diseases of the genito urinary system in both sexes. By DAVID JONES, M.D., &c. Second edition. Edited by Dr. GORDON JONES, surgeon to the Home Hospital for urinary diseases, 10, Dean Street, Soho, London. Simpkin, Marshall & Co.; C. Mitchell & Co. (Red Lion Court, Fleet Street).

patients unsuccessfully who were afterwards successfully cured by the writer were also set out; but to avoid the appearance of individual reproach they were not allocated as they might have been, but only given collectively. Again however the writer falls short of satisfying his professional objectors who are first dissatisfied with the want of means for verification, and then with the best and fullest means that can be offered. They are evidently driven into a corner. They now complain of the indelicacy of publishing the names of professional men. Why then, by taunts and inuendos, did they practically compel it? They throw every doubt that can be suggested on the writer's published records, and next complain of the necessary information afforded for clearing them up. What is it they want? Their requirements are mutually contradictory and cannot be met. The writer therefore relinquishes his efforts to satisfy them, and henceforth will consider only the suffering part of the public with whom he is immediately concerned, and whom alone he will seek to assure that his methods of diagnosis and his means of cure are such as to bear out what he professes to accomplish. Nothing, he would add meanwhile, is more congenial to him than the cultivation of that professional delicacy of which he is severely reminded; but delicacy that is not reciprocal is apt to eat itself away, and his professional friends ought not to be surprised when, in vindicating himself against their too obvious jealousy, he is compelled to disregard it.

As to the propriety of publishing his views and experiences, if that is to be contested the writer will offer only one remark. His exceptionally successful diagnosis as well as the value of his treatment are either true or they are false. If true, the afflicted portion of the public ought clearly to have the fact brought home to them. If false the medical profession at large are bound to inquire into the pretension and expose the error. In either case, where so much is claimed publication is justifiable as a duty. That the profession who are recognised as the guardians of the public interest in all that pertains to health and physical well-being may be facilitated in doing their part in the writer's regard, he herein furnishes materials for examination and criticism, and formally invites a systematic refutation of what he has done in the past and continues to do in the present. Therefore if the reader will pardon the apparent presumption for the sake of the logical accuracy, he categorically formulates his pretensions:

1. That every case he has published is in every essential and material respect *absolutely true*.

2. That he has peculiar means at command of diagnosing stone as well as other diseases of the bladder and prostate ("obscure" as they are called), which high medical and surgical authorities the world over do not possess, and have therefore failed to diagnose, to cure, or even to relieve cases in which he has perfectly succeeded.

3. That, further, in many of the cases he has cured the patients must otherwise have died years ago of stone in the bladder, without any one being cognisant of it had not his timely help been sought.

In confirmation of this last assertion he points particularly to the cases of Captain A. C. CLARK, R.N., of Bombay, of Admiral Sir GEORGE ELLIOT, K.C.B., of the Hon. L. D. E , of the Rev. THOMAS HEATHCOTE (since however deceased), of Mr. DAVID BOWTLE, and Mrs. E. HARSANT, and he could easily add many more. In many of them the success was mainly due to the "spray treatment," which enabled the cause of the trouble to be discovered, and allowed the stones to be removed. In these cited cases, or most of them, in fact the most eminent specialists in England, France, India, and in Germany, had made repeated examinations, and declared that no stone was present, yet the application of the spray treatment soon made it evident they were wrong, and the writer was with little or no difficulty enabled to remove the calculus. In conclusion the writer will only add that, to

Q

establish the fact that his treatment has immense advantages over all other known means, he offers the best evidence in his power, and the cases above cited, taken alone, ought surely to bring home conviction to every unbiassed mind. If indeed the reader would take heed of the pitiful accounts of the patient's sufferings, as given by themselves in coming for the writer's assistance as contrasted with their satisfaction and gratitude on recovery after failure elsewhere, he would cease to regard the frivolous complaints of interested medical men. The objections of the profession are often worse than frivolous. A certain practitioner in Tavistock Square, for example, refused to come and witness an operation by the writer upon that gentleman's own relative, because forsooth, the writer was a homœopath. What immediate connexion is there between homœopathic treatment and operative surgery? And where, the writer may be permitted to ask, is the "professional delicacy?"

No. 48. V. DE M.

Leading Article from the LONDON WEEKLY TELEGRAPH, *August 17th, 1889.—Stone in the bladder and enlargement of the prostate, the stone could not be discovered till after the spray treatment was used.*

Now that Parliament is about to inquire into the management of our hospitals, we would direct the attention of the public to a self-supporting hospital, that is quietly doing a vast amount of good and turning out patients thoroughly cured, many of whom have been pronounced incurable by the most eminent physicians, patients who have tried many of our most popular hospitals, only to be told that they were incurable.

The hospital we refer to is the Homœopathic Home Hospital in Dean-street, Soho, for stone and diseases of the bladder, of which Dr. DAVID JONES, of Welbeck-street, is the founder. His name has now become far and near as a household name for these peculiar diseases, some of the most distressing to which humanity is heir. We speak from personal knowledge when we assert that he has effectually cured thousands of cases, many so difficult that they have baffled the skill of medical practitioners whose names stand highest upon the scroll of fame, men whose decision would be looked upon as final, and whose inability to cure would, in the usual course of things, be the death-knell to all hope. But where these sufferers have been fortunate enough to come to Dr. JONES, he has performed miraculous cures, even upon patients long passed the allotted years of three score and ten. Both men and women who have tried the most popular hospitals and their most skilful doctors and been discharged as incurable, have in a short time been thoroughly cured by Dr. JONES's treatment. He has brought to bear the knowledge and skill of a long life upon the cases he undertakes, with the most satisfactory results. So that there should not be the slightest doubt as to what he affects, he gives in print the names and addresses of the patients he has cured, together with their letters, also the names of popular London hospitals, and the most eminent surgeons and physicians of the day who treated them unsuccessfully, and in many cases pronounced them incurable.

Such astonishing facts are indisputable, and afford indubitable proof of the extraordinary powers possessed by one whom we believe to be the greatest specialist extant. We feel all the more pleasure in penning these lines, knowing what he has done in our own individual case. For a year we were suffering most acute pain, which our West-End physician said must be stone. He examined us, but could find no evidence of stone ; therefore, said it could not exist. The pain still continuing, we went to a large hospital, and another medical man said we had not got stone

After suffering months of pain, we came across a gentlemen that had been cured of a similar disease by Dr. JONES, after having been given up by other doctors. He could scarcely move about when he first saw Dr. JONES, but was soon permanently cured. On our first visit to this eminent specialist, he said he felt assured we had stone, but that he was certain he would soon cure us. After several applications of his spray treatment to reduce the enlargement of the prostate gland, we were placed under ether and a search made for the stone which was found embedded in the bladder. After more spray treatment, the stone was released, then we were placed again under ether, and the stone crushed and brought away painlessly. An instrument containing an electric light was inserted in the bladder to see that it was thoroughly cleared. We were able to do some writing the second day after the operation, and on the third day we left the hospital and went by railway and walked some distance home. We were told that we were a walking miracle. For the first time in a year we were free from pain, and feel no ill effects of the operation. Our case, which seems so extraordinary is only one in many, and we simply cite it to give honour where honour is due, and to let suffering humanity know where it can be treated and restored to health. We need only add that Dr. JONES's invaluable services are within the reach of the poor as well as the rich, and we consider such a man is a public benefactor, whose name and fame should be known wherever there is a sufferer that can be cured by his treatment.

IMPROVED TREATMENT.

Dr. DAVID JONES's treatment improves, of course, under the suggestions of accumulative experience. Medical skill and resources have no finality. The subjoined group of cases serve to show that his methods have advanced since the publication of the last edition of his book. They illustrate in particular, the immediate relief given and cure effected, after prolonged and unsuccessful treatment by medical specialists. The cure of severe cases given up by others, Dr. JONES has been wont to compute by weeks; here it is reckoned by hours :—

———————————

No. 48.

The first case is that of W. M. R., solicitor, from Norfolk, who had been troubled severely with urinary irritation for two years— he had consulted his local medical attendant without relief—after this, he came to a medical friend in town (who had purchased his late father's practice) and he, in turn, sent him to a specialist for urinary diseases, who examined him for stone twice, but getting no relief from his treatment, he suggested an operation, necessitating his keeping in bed for a month, as he believed there was something in his bladder that wanted removing, and added, there might also be a stone in the folds of the bladder, not discovered by sounding.

The writer only saw the patient once, and administered a "spray," requesting him to send a report in three days, as to the effects of the treatment.

Norfolk,
April 8th, 1886.

DEAR SIR,

The three days having elapsed since you treated me, I gladly drop you a line to report the result. I can most truthfully say that I am considerably better, suffer comparatively no pain, and (wonderful to relate) I have slept better for the past three nights, than I have done for quite two years past. Instead of jumping out of bed eight or ten times, I have had three nights sound sleep. Monday and Tuesday nights I got out of bed once, but last night I went to bed about eleven, and never had occasion to get out until about half-past five this morning. This is marvellous for me, and you certainly have given me ease, I hardly dare hope permanently, it would be a miracle if you had, but you have put me in a state of freedom from pain and irritation, that I have long been a stranger to.

With best regards,
Believe me, yours faithfully,
W. M. R.

Mr. W. M. R. again writes after an interval of seven weeks. The "miracle" which he hardly dared to hope for, seems to have been nearly realised, and the freedom from irritation which he experienced the first three days, was but little broken afterwards. As appears from his letter, he is now fairly restored and jubilant. The following letter was an answer to an enquiry after his health :—

Norfolk,
May 29th, 1886.

MY DEAR SIR,

According to your desire, I now drop you a few lines to let you know how I am going on, and have much pleasure in being able to state I am progressing most satisfactorily, so much so that I intend going to Newmarket to-morrow to see the "1000 Guineas." I urinate on an average from five to six times a day, and with very little, sometimes not *any* pain, and my nights are easy : I seldom get up to urinate until about six in the morning, in fact, I have had but one recurrence of pain since I last wrote you ; I attend most strictly to your instructions, the only rule I break is smoking, as the weather has been so horribly cold down here, that I have had my pipe more frequently indoors than out. This is a dreadful country for easterly and north-easterly winds ; we are not more than eleven miles from the sea as the crow flies, and the coast is flat, so that we get them in perfection.

I most sincerely thank you for the interest you have taken in my case, and I will write you from time to time as to how I go on ; should I have a relapse, I shall certainly come up at once to see you. It seems incredible that 1 should have suffered as I did for two years, and that you should have comparatively set me all right with only one application of your marvellous treatment; I am truly grateful to you, and with kind regards,

Believe me, yours faithfully,

W. M. R.

No. 49.

The next case, that of E. H., Esq., more than confirms the immediate success for which Mr. R. was so grateful. Mr. H. was cured from the date of his first visit, after vainly trying other doctors for four months, one of whom was a specialist of high standing.

E. H., Esq., aged 66.

Consulted the writer on June 20th, 1885, giving the following account of his sufferings :—

1. In November, 1884, took a bad cold which affected the neck of the bladder, occasioning intense irritation, and a constant desire to urinate.

2. He was attended by his medical man four months, but getting no better, a specialist noted for bladder diseases was called in consultation, and prescribed various remedies, but without any benefit. Medicines seemed to upset him.

3. Being unable to attend to his professional duties he became anxious and disheartened, when by accident he saw, and purchased the author's book on " Diseases of the Bladder and Prostate," and after reading it, placed himself in his care.

4. In addition to the usual symptoms of inflammation of the bladder—due to prostatic disease—there existed numbness and itching of an unbearable character at the neck of the bladder.

5. The urine was highly alkaline and loaded with fœtid mucus.

6. After emptying the bladder to the best of his ability, there still remained three ounces of residuary urine which he was unable to pass, and which were drawn off by catheter.

The "spray" treatment was administered, and the patient assured that in all probability he would soon be well.

On June 24th, he visited the writer, and said he thought he was really cured, and added, he had had no pain since his former visit, with the exception of a little irritation after the spray.

All his former discomfort had disappeared, and the urine instead of being dirty and full of stringy offensive matter, had become clear and natural. The burning pain had quite gone, and in fact he felt quite himself again.

After nearly a year had elapsed, the patient wrote the following letter, in answer to one sent to enquire after his health :—

Portland Place, W.

15—3—86.

DEAR SIR,

It affords me much pleasure to be able to testify to your successful treatment of my case. I was under two doctors for about four months, one of whom is of very high standing in the profession, and a specialist for urinary diseases. I took eight different kinds of medicine without any improvement, indeed they quite upset my general health.

Happening to see an advertisement of your book in a newspaper, I procured a copy, and after perusal came to you. From the *first* visit up to the present time, I have felt no pain or inconvenience whatever. I have taken wine and spirits in moderation, but have had no return of my painful disease. My friends tell me (and I believe myself) that as it is now twelve months since I came to you, I may conclude I am perfectly and permanently cured. I shall be very pleased to see anyone requiring confirmation in respect to your treatment of my case, and with heartfelt thanks to you.

I remain, dear sir,

Yours faithfully,

E. H.

E. H. continued well up to June, 1890, and writes,—

DEAR DOCTOR JONES,   June 5th, 1890.

It has given me the greatest pleasure to testify to Mr. P—— of your successful treatment of my case, and the straightforward manner in every way of your dealing with it : fancy being told in 1886 I would never be better, and be a sufferer all my life ! ! ! —I have been well ever since you cured me.

E. H.

No. 50.

A third case, that of Mr. HOLMES, of Cambridge, shows the perfection of the spray treatment, two applications of which brought a derelict sufferer back to the full satisfaction of life. Cambridge has a high repute for medical skill, and is a centre of medical teaching; but five months of its best physicians and surgeons in this case failed to afford relief. This patient regards the promptness of Dr. JONES' cure as marvellous :—

44, Panton Street,
Cambridge,
May 31st, 1886.

MY DEAR SIR,

I intended writing to you sooner, but thought it better to wait awhile after such speedy cure as you effected on me, in case of a relapse of my painful disease. I am thankful to say I continue quite well, and have no pain or inconvenience of any kind. Instead of having to get out of bed 12 or 14 times every night to pass water, which I did with severe forcing and scalding pain, I only get up once or twice to urinate with perfect ease and freedom.

I consulted three of the first physicians in Cambridge, and a homœopathic physician, and was under their care for five months without deriving the slightest benefit. Before I came to you, my doctors (very eminent ones too) said all that could be done for me had been done, so I gave myself up to fate, and asked them not to come again. Soon afterwards I saw an advertisement of your " Home Hospital" for stone, &c., in the " Christian World," and decided to come to you. Strange as it may appear, you did for me with two applications of your spray, what my other doctors could not do in five months. Life is now a pleasure to me; I eat well, and take stimulants (which I was forbidden to do before) without any discomfort. I am truly thankful I ever came to you. My friends say my cure is nothing less than a miracle. I shall be very pleased for you to make any use you like of my name and case. I make it my duty to talk to all I meet about your successful cure.

I am, yours gratefully,

WM. HOLMES.

June, 1890, Mr. HOLMES writes, " Thank God I am perfectly well."

No. 51.

Disease of the prostate gland, and severe inflammation of the bladder of several years' standing, cured by the administration of three "sprays."

Mr. WILLIAM REYNOLDS, aged 75, Barrington, Cambridgeshire.

Consulted the writer, accompanied by his son, on July 12th, 1886, and stated that he had been suffering from weakness of the bladder for many years, but as the inconvenience was not very great, he attributed it to advancing age. Three years ago, however, he was seized with complete stoppage of the urine, and was obliged to send for his local doctor, who gave him relief by using a catheter and drawing off about a quart of water. The same treatment had to be adopted on the following day, and he was then advised to go into the Cambridge Hospital. He did so, and was there for a week, when he was discharged, and told to continue using a catheter himself. He followed out this advice for more than a year, but the catheter set up such pain and soreness that he was compelled to leave it off. After that he got gradually worse, and for more than a year and a half before his visit to the writer, he had been obliged to urinate every ten or fifteen minutes day and night (generally worse at night), the act being attended by intense scalding and straining, the pain being more severe at the anterior part of the perinæum, and just above the scrotum.

Occasionally during the daytime there would be longer intervals of relief, but when such was the case, they would invariably be followed by severer paroxysms than usual.

The patient had heard of the successful case of Mr. W. HOLMES, 44, Panton Street, Cambridge (the case immediately preceding this one), and was on that account advised to try the "spray treatment" himself.

An examination satisfied the writer that he had the usual form of prostatic enlargement, accompanied by severe "cystitis"

APPENDIX.

NAMES AND ADDRESSES OF PATIENTS.

APPENDIX.

Names and Addresses of Patients whose Cases are described in Part V.* of this Edition.

No. of Case.	Page.	Initials.	Full Name.	Addresses.
1	68	E. S.	Mr. Edwin Stevens.	97, High Street, Hounslow.
2	72	T. L.	Mr. Thomas Ludlow.	16, The Grove, Crouch End, Hornsey.
3	78	W. R.	Mr. William Ryder.	(Solicitor retired). 324, Brixton Road, Surrey.
4	84	Sir F. H.	Gen. Sir Frederick Horn.	Buckby Hall, Rugby.
5	86	Canon C.	The Rev. Canon Cockin.	
6	91	W. F. L.	Mr. William Lacey.	
7	95	E. B.	Mr. E. Bentall.	
unnumbered?	99	I. C. W.	Mr. I. C. W. Ibbs.	Northamptonshire.
8	100	R. C.	Mr. Robert Cole.	Stanwell, near Staines.
9	103	I. B.	Mr. I. Brittain.	Kentish Town.
10	106	W. P.		
11	109	B. H.	Mr. Beaman.	New North Road.
12	111	J. A.		
13	115	G. H.	Mr. George Hussey.	44, Manchester Street, Southampton.
14	117	G. G.	Mr. George Garnett.	(Messrs. Garnett & Sons), Leeds.
15	118	G. F. V.	Mr. George F. Vallins.	5, Manor Street, King's Road, Chelsea.

No. of Case.	Page.	Initials.	Full Names	Addresses.
16	119	J. P.	Mr. James Pink.	Fareham, Hants. *
17	124	G. S.	Mr. George Sowerby.	c/o Mr. Tyson P. Doyle, 17, Blencowe Street, Carlis'e.
18	131	E. T.		
19	136	F. L.	Mr. Frederick Lawley.	122, Fairbridge Road, Holloway
20	138	J. B.		
21	140	E. O.		
22	142	J. B.		
23	143	J. H.	Mr. J. Hutchinson.	26, Winterwell Road, West Melton, Rotherham.
24	145	J. A.	Mr. James Adams.	
25	146	S. H. T.	Mr. Samuel H. Tonks.	South Yorkshire Asylum, Wadsley, near Sheffield.
26	152	S. M.	Mr. Stephen Mann.	Woodgate End, Epsom.
27	154	H. D.	Mr. Henry Daykin.	23, Albert Street, Barnsbury.
28	158	Mrs. R.	Madame Reymond.	Vers de Lac Sentier, Val de Young-Vaud, Switzerland
29	161	E. R.	Mrs. Kosindell.	32, Bedford Row, Holborn.
30	163	S. M.	Miss S. Munns.	
31	165	E. M. A.	Mrs. E. M. Allen.	Late of Religious Tract Society, 164, Piccadilly.
32	167	S. C.	Mrs. S. Cookman.	37, Maryland Road, Harrow Road
33	170	F. T.	Mrs. F. Tookey.	Greenfield, Ampthill, Beds.

* In every instance where the full Name and Address are given, those Patients may (by permission) be personally communicated with by anyone desiring information respecting their Cases and in almost every other instance the writer is authorized to furnish such information as he may think proper.

No. of Case.	Page.	Initials.	Full Name.	Addresses.
34	174	E. H.	Mrs. E. Harsant.	
35	179	M. C.	Mrs. Cairns.	32, Abbeyfield Road, South Bermondsey, S.E.
36	185	Mrs. S.		
37	189	Mrs. E. A.		27, Coburn Street, Bow Road.
38	191	Mrs. T.	Mrs. Tolly.	
39	193	A. L.	Mrs. Lewis.	19, Cranbury Place, Southampton.
40	195	Mrs. E. L.	Mrs. Dossiter.	
41	197	E. D.	Mrs. Smith.	
42	200	S. S.		
43	202	Hon. L. D. E.	Mr. David Howtle.	36, Sidney Street, York Road, King's Cross.
44	206	D. B.	Mr. Henry J. Barrett.	8, Clyde Terrace, Anlaby Road, Hull.
45	209	H. J. B.		Bristol.
46	211	O. A. R.	Case of a valuable Cob the property of C. G. Duff, Esq.	1, Lennox Gardens, S.W.
47	212			

IMPROVED TREATMENT.

No. of Case.	Page.	Initials.	Full Name.	Addresses.
48	229	W. M. R.	Mr. W. M. Rumbelow (Solicitor).	Fakenham, Norfolk.
49	231	E. H.	Mr. Edward Hodges.	31, Devonshire Street, Portland Place.
50	233	W. H.	Mr. Wm. Holmes.	44, Panton Street, Cambridge.
51	235	W. R.	Mr. Wm. Reynolds. London reference (his son) Mr. J. Reynolds.	Barrington, Cambridgeshire. 2, Haverstock Hill.

APPENDIX.

Names and Addresses of Patients whose Cases are reported in the Second Edition of Urinary Diseases. Edited by GORDON GRIFFITHS JONES, Surgeon to Home Hospital for Stone, Soho.

No. of Case.	Page.	Initials.	Full Name.	Addresses.
1	1	Capt. A. C. C.	Capt. A. C. Clark, R.N.	Lloyd's Surveyor, Bombay.
2	4	Adml. Sir G. E.	Admiral Sir George Elliot, K.C.B., &c.	6, Castletown Road, West Kensington.
3	7	H. H. U.		
4	12	C. B. D.	Mr. Clement B. Dixon.	Crosslea House, Henley Road, Ipswich.
5	16	S. P.	Mr. Stephen Parker.	25, Drapper Street, Bermondsey.
6	21	T. D.	Mr. Thomas Day.	Resided Mar. 2, 1882, at 39, Swinton St., King's Cross Rd.
7	23	Rev. C. G. S.	The Rev. C. G. Squirrell.	Stretton-under-Fosse, Rugby.
8	24	T. N.	Mr. Thomas Newstead.	Resided, Feb. 2, 1883, at 132, Weymouth Ter., Hackney Rd.
9	25	Rev. J. B.	The Rev. John Breese.	
10	28	W. C. L.	Mr. William C. Lay.	Newbury, Berks.
11	30	C. W.	Mr. Charles Wilson.	Garstang, Lancashire.
†12	33	J. M.	Mr. John Moore.	23, Egerton Street, Chester.
13	37	Rev. T. H.	The late Rev. Th. Heathcote.	Lenton Vicarage, Lincoln.
14	39	J. B.	Mr. Joseph Bell.	(Reference to case) The Rev. J. Brennan, Grays, Essex.
15	43	J. R. E.	Mr. J. R. Edwards.	72, Frederick Street, Grays Inn Road.
16	46	W. G. M.	Mr. W. G. Murless.	131, Shakespere Road, Hackney.

No. of Case.	Page.	Initials.	Full Name.	Addresses.
17	48	J. J. C.	Mr. J. J. Collins.	Lordship Lane, Forest Hill.
18	49	C. S.	Mr. C. Sutcliffe.	Trent College, near Nottingham.
19	50	T. S.	Mr. Thomas Stevens.	
20	52	W. C. R.	(Reference to case).—	The Rev. J. Richardson, Stretton-on-Dunsmore, Rugby.
21	54	W. R.	Mr. Walter Robson.	16, Stanbo Lane, Boston, Lincoln.
†22	56	W. M. R.	Mr. W. M. Rumbelow.	Fakenham, Norfolk.
23	57	E. H.	Mr. E. Hodges.	31, Devonshire Street, Portland Place, London.
24	59		Mr. William Holmes.	44, Panton Street, Cambridge.
25	61		Mr. William Reynolds.	Barrington, Cambridgeshire. London reference, Mr.
26	63			Reynolds, 2, Haverstock Hill.
27	65	M. J.	Mrs. M. Hughes.	3, Broomgrove Road, Stockwell (Mother of patient).
28	69	E. M.	Minnie Jacobi.	
29	70	S. B.	Mrs. S. Bown.	386, Albany Road, Camberwell.
30	72	H. H.	Miss Harriet Hudson.	
31	75	H. E. A.	The late Miss H. E. Abbott.	14, Imperial Buildings, Cheltenham.
32	77		(Reference to case). -	Miss Marion Hutton, 20, Adam St, Manchester Square.
33	81	L. A.	(Reference to case).—	The Matron, Home Hospital, 10, Dean Street, Soho.

* In every instance where the full Name and Address are given, those Patients may (by permission) be personally communicated with by anyone desiring information respecting their Cases and in almost every other instance the writer is authorized to furnish such information as he may think proper.

† N.B.—These Cases have had a slight return of irritation, but are now cured.